Collective Consciousness of Afrocentric Intellectuals vol 1

© Kofi Piesie Research Team Same Tree Different Branch

Kofi Piesie/Mossi Warrior Clan
Copyright 2020 by Kofi Piesie Research Team

All right reserved. No Part of this book may be reproduced or transmitted in any form or by any means, electronic or mechanical, including photocopying, recordings, or by any information storage and retrieval systems without the written permission of the publisher.

Printed in the United States of America

Table of Contents

Introduction 5

Chapter 1: Collective Consciousness The Multifaceted Motivation of a Self directed learner... 9

Chapter 2: Fall of The Blackman: Caught In The System... 22

Chapter 3: Can your freedom of speech be used against you?..................................... 37

Chapter 4: Towards a deeper understanding of Indigenous African Information Systems as a catalyst for ideological, socio-economic, and holistic liberation.................................. 51

Chapter 5: Assessing Emotional Intelligence and its Significance to African Americans..... 75

Chapter 6: Ethiopianism, Ethiopian Church & AME Church.. 99

Chapter 7: Examining available data on the Pros and cons of eBooks versus physical paperback books................................... 118

Chapter 8: An interesting Message from AI... 133

Chapter 9: Introduction into Authorship.... 140

INTRODUCTION

This publication was created with specifically the black youth in mind. But just because the intended target is black youth, this doesn't eliminate the possibility of any other demographic from reading this and obtaining some type of edification. Each author of this publication is African American, and our objective was to attempt to document information that we felt was little known or not dealt with in an in-depth matter that would be beneficial to the reader of any age, ethnicity, or background. If you seek to obtain knowledge, then this book is for you. I just wanted to make the target audience we were writing to be black youth so we could be more meticulous with what we chose to convey that young minds are so impressionable. Also, they are our future, so why not create a publication with them as the focus that will assist them in this journey, we call life? So many times, in my journey, I ran across information that I reflect on and wish someone would have told me earlier.

Because of this, I was inspired to put these types of things in book form and get some of the brightest and sharpest minds to help out by following my lead in documenting things they thought would help enlighten our youth and knowledge seekers in general. So based on this objective alone, this publication cannot fail by default. It is a victory to the authors on the strength that it is no longer a thought or talk we are reaching a stage of reality where things are in our hands all we have to do is manifest. This is accomplished by turning thought into action, and that is exactly what each of these authors under the umbrella of the name Tep Heseb Warriors mission was. I define what a /tp Hsb aHAw/ was in my 2022 publication Religious Beliefs Revisited the tep heseb edition.

" Another phrase that may have captured the reader's attention on the cover that many aren't familiar with is "tep heseb edition." You may have visited my Facebook profile and seen /tp hsb ahaw/ and had no clue what that meant or what it was. I have heard everything from people thinking I changed my name to people assuming it's an acronym that they just don't know what the letters represent for a gang or click. The reality is I translate exactly what it means under my picture: /tp hsb/ (Correct method) /ahaw/ (Warrior). It isn't a name given to me by any group or organization; it is honestly a title I created and chose to embody, influenced by innovative ancient AFRIKAN genius always to be a reminder that the idea of white man science is fallacious. " (McCray. C,

2022). This Afrikan genius is represented in each individual who was hand-picked for this publication, and I am honored they participated in the development of this project and hope it accomplishes its mission, and our objective is reached. We appreciate your purchase and the time you are taking to seek edification through our book. It is a tool we offer to assist in your "Innovation Over Ignorance" while still embodying Kofi's "As I learn, We all Learn" moniker. Tap into the "Collective Consciousness of Afrocentric Intellectuals."

Chavis "Tp-hsb Ahaw" McCray of KPRT and Creator of the Tep heseb Warriors as well as the Innovation Over Ignorance Inc brand

CHAPTER ONE

Collective Consciousness The Multifaceted Motivation of a Self-directed learner

Collective Consciousness The multifaceted motivation of a Self-directed learner

Chavis Tp hsbAhaw McCray

This chapter was influenced by an essay I had to submit for an assignment in an online class I had to take at the online school I attend called the University of the People. The assignment and class, in general, were designed, in my opinion, to educate the student about the pros and cons of the online learning environment, essentially having us dig into the reality of what type of student thrives in this type of environment and what type fails. The fact that this online learning thing was new to me and the fact I had reached this stage in my life where knowledge was like a drug I was addicted to me, I immediately had an epiphany doing the assignment that this topic was multilayered and was essentially applicable to everyday life and would make the perfect topic to explore in depth further and share in a publication. Yes, while attending this online school, I was also already a part of a research team I'll refer to as KPRT (KOFI PIESIE RESEARCH TEAM) who I previously was only a fan and supporter of for the past few years as I heavily consumed 100s of hours of their content as they were the most reliable and objective source of African information my algorithm ever came across. These gentlemen had unmatched character and an obvious love for edifying their people on the subject of African culture to depths only the most serious researcher and lover of all things can really and truly appreciate and understand. These brothers resonated with me because of this and I could see they were solid in their studies and their intent was pure as they were so consistent in providing a plethora of insight on some of the less talked about topics in great detail in regards to African Culture and worldview. For me it started with MOSSI warrior Clan and the Seshew maa my medew netcher channels along with Houston's finest Asar Imhotep channel that I stayed tuned into, supporting, and participating in via social media. The individuals on these platforms were different from why I had subscribed for years prior. They were serious about learning and correcting misconceptions and misinformation dealing with the African diaspora as a whole. They also challenge each other to a standard of methodical approach and rigor vs opinionated pseudo-nonsense. One channel you could usually find everyone hashing out this pseudo nonsense was my good brother Bobby Baltimore Banga aka Ankh West channel the Real Black Atheist YouTube channel. Something happened to where I believe Bro BOBBY (inside joke) lost his channel for while and he ended up creating the Pseudo Killer channel on YouTube. Bro Ankh since I first started following his wave was always known as the bad guy who beat misinformation for years from some of the craziest characters on social media. He was also what coined an RBA (REAL BLACK ATHEIST) that was a major advocate of SCIENTIFIC LITERACY and calling out pseudo-scientific claims debunking them with raw unpopular truth. I loved it and when I found out he wrote a book which some may call a pamphlet defining this concept he labeled himself I was immediately inspired. Ankh was from the streets he was an open book and as straight up as you could find him and will argue with literally anyone. He was a real one in my book from the start and is the piece that connects everyone. He always gave his people a platform to be able to participate or at the least view these intelligent discussions on everything from religion to evolution to vaccines. He was a major advocate of "source up or shut up" that terrorized and intellectually thrashed pseudo social media personalities, quacks, and the "Conscious Community" as a whole. He assembled the Amen Ra Squad who intellectually bullied conscious community pseudos who lack basic research methodology and struggle in the area of scientific literacy. He had a hand in assembling and is a founding member of the MOSSI warrior Clan and creator of the pseudo killers. His book resonated with me after reading it cuz I respect the fact that he was taking ownership of this label he created and wouldn't allow anyone to place him in the box with another atheist by taking his pen and putting it to paper to document it and embody it for eternity. That whole thing was exactly how I felt. I didn't want to let anyone put me in a box and I took a position I wanted to document that ended up publishing later down the line and was able to give to him directly. I was able to express my gratitude for all the information he helped correct and tell him how much he influenced me. I was able to receive praise from him for essentially developing an intellectual artillery that he felt was above the average individual in the streets and joining the battle for our people's minds. He told me I was the future and was different and he has been a major inspiration along with others such as Wudjau, Asar, KPRT, AND THE MOSSI to embrace Kofi PPiesie's "As I learn, we all learn" motto and title of his book one should obtain and consume if you don't have you need to get for an understanding on why we all do what we do and have this mission to advance our people mentally through proper Edification via publication. Which is what this publication's mission is as well. The topic I'm choosing to facilitate this objective to try and tie in what began this chapter is the subject of collective and individual multifaceted motivation of self-directed students of life in the African diaspora. I aim to take what I learned from this in depth study and attempt to make it applicable to the life of today's "blacks" and offer an understanding of something I believe isn't thought about or talked about much in regards to the intrinsic and extrinsic motivation of our people and how if we collectively identify and approach this subject from the same way organizations who study these dynamics for optimum performance and productivity for their organizations progress we could achieve the same types of success just in a different way. Instead of from a random white organization, we view Africans across the diaspora as the organization and each one of us as employees/ students/ representatives of the "organization" striving to be able to offer necessary performance and productivity in all significant areas of life to not only maintain but advance the "organization".

One of the first things that caught my attention in this essay I submitted that I felt the need to touch on was that a self-directed learner "Defined by adult education expert Malcolm Knowles, self-directed learning describes a process by which individuals take the initiative, with or without the assistance of others, in diagnosing their learning needs, formulating learning goals, identifying human and material resources for learning, and evaluating learning outcomes.". (Briggs, 2015)

I thought to myself what if every black person I knew understood and took this initiative seriously? Where would we be collectively if we challenge ourselves with or without the assistance of others to diagnose our learning needs, formulated what wanted to achieve concerning learning goals, and made a regular conscious effort to recognize our resources for learning, and gauge those learning outcomes? What if we were all scientifically literate and approached life from an angle that wasn't so deeply rooted in faith, indoctrination, and religious fundamentalism? What if we didn't have so many people who were willingly ignorant in our "organization"? What if critical thinking and scientific reasoning were the norms? Would we still be the labor class that we are in the United States? I am of the position that change starts in the mind first and this type of change mentally could trigger a change in our current situation leading to advancement and progression.

The essential aspects of self-directed learning as Briggs accurately describes are "Playfulness, Autonomy, Internalized Evaluations, Openness to Experience, Intrinsic Motivation, Self-Acceptance, and Flexibility". (Briggs, 2015) Playfulness describes the curiosity and the creative aspect while autonomy speaks to an individual's agency in their learning. All of the characteristics she mentions are self-explanatory and paint a vivid picture of what fuels this particular type of learning that is necessary to grasp and embody the mindset you need to identify as self-directed in your learning. Research indicates SDT (Self Determination Theory) is a theory that "suggests that human actions, such as creative and innovative performance, are strongly affected by the type of underlying motivation and are triggered by individual motives and needs. According to the SDT, motivation varies along a continuum between controlled and autonomous motivation." (Ryan and Deci, 2000). This would apply at work, in school, on social media, at home, etc etc. So basically how something motivates you determines interest and that interest is triggered by individual motives and needs. If you have no motive or need for an understanding of linguistics you more than likely will not have an interest in linguistic discussion or subject matter. The same thing pretty much applies across the board whether we are dealing with virology, immunology, vaccinology, sociology, psychology, physics, astronomy, basic math, English, finances, the economy, government, law, agriculture, engineering, and so on. The average black person in my opinion thinks we don't even use the stuff we learned in school so what need would they have to be motivated to have an interest in all of these things? Well, the answer would more than likely be they don't have a need and would explain our lack of representation in those fields and why we are constantly disseminating misinformation in those areas as a collective. This was exposed with the pandemic and how susceptible we were to pseudo-claims about the covid vaccine and the virus itself. A lot of us probably still don't know sars cov 2 is the virus that causes covid 19 the disease or that it ain't no food you can eat that will cure it or protect you from it. We still have blacks who think those who got vaccinated will turn into zombies or will drop dead in 10 years not Knowing there has never been a vaccine that had such an outcome in the history of vaccines. This type of reasoning is grounded in willful ignorance and intellectual laziness influenced by a lack of intrinsic motivation. "Intrinsic motivation describes the undertaking of an activity for its inherent satisfaction.." (Nickerson, 2021) Being intrinsically motivated about the subject would make it oxymoronic to be willfully ignorant and intellectually lazy logically. You would want to know and challenge yourself outta pure curiosity whether vaccines were killing people or not. You would think twice about whether ineffective claims of the vaccine were true and you would know how to verify that. You wouldn't parrot social media memes as fact. I wouldn't even care for a lack of extrinsic motivation as well. There is simply no need outside of fear-based motivation to see a need to have an interest.

If a person believes there is nothing to fear because the virus is fake or it's a plandemic that belief influences their behavior and technically eliminates that fear. Which is counterproductive on a collective and individual scale. "Extrinsic motivation is a construct that pertains to whenever an activity is done to attain some separable outcome. Extrinsic motivation thus contrasts with intrinsic motivation, which refers to doing an activity simply for the enjoyment of the activity itself, rather than its instrumental value." (Nickerson, 2021)

Out of the 2 types of motivation intrinsic has been identified as being associated with increased performance. "Performance, productivity, and sheer enjoyment of work have all been found to be greater in people with higher levels of intrinsic motivation." (Ben-hur & Kinley, 2016) "Research shows that for more complex tasks, intrinsic motivation helps to drive higher performance. When the job itself is enjoyable and interesting, employees work harder and with greater focus and commitment without the need for extrinsic motivators." (Turner, 2017) If we view living as a job and we are intrinsically motivated then our health should be our focus and commitment we work hard to maintain by knowing the things that will help you lose your "job" fast and the things that extend your "employment". This isn't an attempt to minimize the relevance of extrinsic motivation as research " has posited two types of motivation theories. Dualistic theories divide motivation into two types: intrinsic and extrinsic. Multifaceted theories, in contrast, recognize several genetically distinct motives. Intrinsic-extrinsic dualism fails on at least three counts: construct validity, measurement reliability, and experimental control. Many researchers have thus moved beyond the study of intrinsic-extrinsic motivation and validated multifaceted theories. When teaching students about the multifaceted nature of motivation, teachers can take several steps to improve their student's understanding of this understudied area of psychology." (Reiss, 2012)

A more up-to-date research has found essentially that "Extrinsic factors can be seen as synergistic extrinsic motivators when they have a positive effect on the outcome" (Fischer, Malycha, Schafmann, 2019) The term synergism is defined as "the interaction of discrete agencies (such as industrial firms), agents (such as drugs), or conditions such that the total effect is greater than the sum of the individual effects." (Merriam Webster, 2022)

The Study investigated "synergistic extrinsic motivators that organizations can use to foster creativity and innovation of their intrinsically motivated knowledge workers" (Fischer, Malycha, Schafmann, 2019) Think of ways this information could be applied and tap into this better understanding of how your motivation is tied to success.

CHAPTER TWO

Fall of The Blackman:Caught In The System

FALL OF THE BLACKMAN:CAUGHT IN THE SYSTEM.

Brother Lavelle

From Slavery to Jim Crow to Mass Incarceration is the destruction of the Black man. In this essay, I will examine these three components that have affected the Black man in the "system." For me to bring out my viewpoint, along with the research that I have obtained, it would be only fair for me to explain the system in three sections.

Beginning with slavery, which is the core of the American prison system. This would be the first stepping stone in demolishing the Black man. I will be going in-depth on the early coalition of Europeans to enslave African families; the Portuguese are the first to ever put Africans in chains. This essay goes out to my brothers who keep finding themselves, in the system. The essay is not to make no excuses, on why the Black man continues to plunge into this trap, but it is to explain why this net/trap is set up for us as men to fall.

To be free from this awful snare, we as men must avoid plunging into these traps created by our oppressors, since the time they intruded on Africa. The more we tumble into these pitfalls, the more we prolong our enslavement. I need the reader to realize, that at one point I found myself caught up in the system. I am quite sure that the reader may know someone that has been caught by the system too.

Another part of this system that I will examine is Jim Crow, rules which happened after slavery. And used against the Black family, particularly the Black man. The consequences for bending or fully breaking these requirements are prison terms and lynching.

Mass incarceration is the last and modern aspect that we endure as a people. Mass incarceration, in my viewpoint, is a mixture of the first two traps, plus an extraordinary stipulation to putting the Black man in complete modern-day slavery.

SLAVERY

When did slavery happen?

As I expressed in my brief introduction, slavery began with the Portuguese who raided the western shores of Africa, around the mid-1400s with an individual named Prince Henry better known as Henry the Navigator. Born on March 4, 1394, in Porto, Portugal, to the parents of John 1 and Philippa of Lancaster. His father John 1 was king from 1385 till his demise in 1433. Henry the Navigator is known for the age of discovery by certain people, but he was also one of the main characters in the Trans-Atlantic Slave Trade. "But his efforts also began the process of European colonization, capitalism, and, ultimately, the transatlantic slave trade." (https://education.nationalgeographic.org)

Many slaves from this period did not see the North American shores but landed in South America. This account would be the first instant that the Black man and his family experienced any sort of confinement. Our people were being abducted and stripped of their nationality, and religion, (KIMOYO), as they would stroll for miles, preparing to be harbored and transported. These were holding tanks, (cells) constructed in the year 1482, and it is dubbed the Elmina Castle. Underneath our ancestors would live in the nastiest constraints, just like prisons today. (https://www.pbs.org/ Elmina)

Our history being carried to the western hemispheres is two-fold, I briefly declared the first fold, and now I want to dive into the North Americas. Even though Africans have been coming to the Americas as slaves mid-1400s (1444) and early 1500s along with Portuguese ships. The first recorded slave anchor to ever hit North America was in 1619 on the shores of Jamestown, Virginia. Lerone Bennett, Jr. lets us know how this stolen ship ended up in the northern part of the Americas. "No one knows for sure. The captain "ptended," John Rolfe noted, that he was in great need of food and offered to exchange his human cargo for "victualle." The deal was arranged. Antoney, Isabella, Pedro and seventeen other Africans stepped ashore in 1619." (Bennett, Jr: 1963: Before The Mayflower)

This act was the first part of slavery but not really. When the first group of Black people exited that ship, they were not slaves but worked side by side with indentured whites, and even enjoyed the same privileges, as their white counterparts. The first Black settlers owned land, voted, and even testified in courts. They even gathered around White people among their communities. That is including living together without forcing people to separate.

These obstacles that our ancestors engrained deeply inside of us, from the dungeons of the Elmina Castle which represents processing including the way they fed them, until the time they were placed on the ships which represents the paddy wagon, and then to the auction block that can be identified as the court system, and then to the auction block what we can identify as the penitentiary. As Black men we must overcome this crime that was upon us.

Slavery was just the seed of this corruption, that we continue to face to this very day. This root of the system has put a curse on the Black man he wished he never experienced. Slavery caused six actions that destroyed us as Black men. 1. Kidnapped from his home. 2. Destroyed him mentally. 3. Stripped his spirituality (KIMOYO). 4. Effeminized him. 5. Gave him laws to abide by. 6. Split the African home.

Before I end this section, let me discuss slave codes, and how they used them against Africans. New York Slave Revolt, started on April 6, 1712, when a group of Africans set fire to a building, on Maiden Lane close to Broadway. Colonists tried to put the fire out, but Africans that were carrying guns, hatchets, and swords took charge of the colonist and they ran off later they would be captured.

Nine whites were shot, stabbed, and some beaten to death and another six were wounded. Twenty-seven slaves were captured six committed suicide the rest of them were executed and burned alive. White New Yorkers were highly upset about this Black insurrection that promoted Black unity, they had to draw out certain laws that were harsher and stricter than the earlier ones.

Here are certain laws that they enforced. 1. Slaves could not meet in groups larger than three. 2. Masters can punish their slaves for anything. 3. Slaves could not obtain firearms if caught they were given twenty lashes. 4. Any slave caught gambling they were whipped on the spot. 5. Any involvement to conspire against your masters would result in immediate execution. (https://en.m.wikipedia.org/NewYorkSlaverevolt)

JIM CROW

What is Jim Crow?

According to the BLACK'S LAW DICTIONARY, "Jim Crow Law. (1891) Hist. A law enacted or purposely interpreted to discriminate against Black people, such as a law requiring separate restrooms for Black people and whites. Jim Crow laws are unconstitutional under the 14th Amendment." (Black's Law Dictionary 5th edition: 432:2016)

Around the early 1860s was the creation of Jim Crow, which of course birthed by White men. They would dress up as Black faced characters and portray our people as slow, lazy, dumb niggers. Jim Crow is also a form of RACIAL DISCRIMINATION that is rooted in the first stage of the Black man's downfall… Slavery….

This would be the time of the Civil War, and this Black-face minstrel would be prominent amongst Whites. So, it would only be right for me to bring up the Reconstruction Era, I would like to quote from Michelle Alexander's book The New Jim Crow to begin my premise about the laws that they put on our ancestors. "The backlash against the gains of African Americans in the Reconstruction Era was swift and severe. As African Americans obtained political power and began the long march toward greater social and economic equality, whites reacted with panic and outrage. Southern conservatives vowed to reverse Reconstruction and sought the abolition of the Freedman's Bureau and all political instrumentalities designed to secure Negro supremacy." (Alexander: pg30: The New Jim Crow)

RECONSTRUCTION BEGAN AND SO DID THE KLAN:

Conservative whites were enraged because of the remarkable success of Black people, who were their enslavers are now passing them up, and putting black footprints that was unbearable to these conservatives. Those conservatives were your old-fashioned Democrats, it is a proven fact that Black people in the times of 1865-1877 (Reconstruction Era) were Republicans, and unlike today the racist bigots were Democrats. As the Reconstruction Era began so did the Klan, which was formed by six Confederate soldiers, but the first Grand Wizard of the Klan was Nathan Beckford Forest.

Nathan Beckford Forest oversaw the massacre of the Black union troops that were stationed at Ft. Pillow, Tennessee, in April of 1864. He became a millionaire by dealing with livestock, real estate, and planting cotton, but his main prize ticket was selling slaves. (https://www.britannica.com/NathanBedfordForrest)

13th Amendment

"The 13th Amendment was proposed on January 31, 1865, and ratified on December 6, 1865. Section 1. Neither slave nor involuntary servitude except as a punishment for crime whereof the party shall have been duly convicted shall exist within the United States or any place subject to their jurisdiction." (Jordan, Terry:pg49: The U.S. Constitution) The 13th,14th, and 15th Amendments are the only three Amendments to ever pass in the Reconstruction Era

Plessy v. Ferguson (1896)

"Jim Crow laws are constitutional under the doctrine of Separate but equal" (Jordan, Terry:pg82: The U.S. Constitution)

In Louisiana police arrested Homer Plessy for refusing to leave a railroad car that prohibited "colored" people. Under Louisiana law, Plessy considered to be "colored," because he was one-eighth of a Black man. The court ruled that the race-based Jim Crow laws did not violate the Constitution if the states proffered separate, but equal treatment; this set up the Jim Crow laws throughout the south. "The dispute arose as a test case to challenge a statute; an example of the Jim Crow laws then being passed in the South as whites sought to embellish their control of state governments." (Hall, Kermit:pg637: The Oxford Companion To The Supreme Court)

How did Jim Crow laws affect the Black man?

Black men and their families were treated equal in certain situations but separate in many of them. At this time laws have hindered us, but this made the Black community stronger and united. From 1890-1965, Black people were still under these stipulations and mistreated. When dealing with each other during the Jim Crow laws era brought out some good in our community, as Black businesses were flourishing.

"Of course the earlier system of racialized social control-slavery- had also been regarded as final, sane, and permanent by its supporters. Like the earlier system, Jim Crow seemed "natural," and it became difficult to remember that alternative paths were not only available at one time, but nearly embraced" (Alexander, Michelle: pg32: The New Jim Crow)

What ended Jim Crow?

On July 2, 1964, President Lyndon B. Johnson, signed the Civil Rights Bill that outlawed discrimination based on race, color, religion, sex, and national origin. (The Civil Rights Act of 1964 formally dismantled the Jim Crow system of discrimination in public accommodations, employment, voting, education, and federally financed activities." (Alexander, Michelle: pg32: The New Jim Crow)

As Black people were beginning to start a new phase in America, something halted our ancestors. Lynching was considered the death penalty as being Black was an actual crime. The Black man quickly adapted to his situation and stayed the dominant force in his household, community, and churches.

MASS INCARCERATION

"While the prison population and black male incarceration, began its historic rise in 1973, clearly the antecedents of that moment were long in the making. A century-long history of brutal racism beginning with slavery and progressing through Jim Crow in all it is permutations throughout the nation set the stage for a modern-day version of oppression in response to developing social and economic conditions in American society." (Davis, Angela J: pg33: Policing The Blackman)

In June 1971, President Richard Nixon declared the War on Drugs as public enemy number one. He also enhanced the federal funding for drug control; this would be the trap that the Black man has faced to this very day. (https://en.m.wikipedia.org/warondrugs)

This action led to the imprisonment of the Black man and is the last stage of this corrupted system; it is not the last stage of the War on Drugs.

On July 1, 1973, Richard Nixon creates the D.E.A. which was designed to fight against a global War on Drugs. The D.E.A. is a combination of the Bureau of Narcotics and dangerous drugs, the office of Drug Abuse Law Enforcement, and different labels. This action is still planned to put an all-out war on the Black man and the community.

On October 14, 1982, President Ronald Reagan followed in Nixon's footsteps and did his "War on Drugs," which he considered a threat to the United States National Security. This was also a dangerous time for the Black community as the "crack epidemic" launched, destroying the lives of the Black man and his family and opened the door for Mass Incarceration. "Reagan greatly expanded the reach of the drug war and his focus on criminal punishment over treatment led to a massive increase in incarcerations for nonviolent offenders." (https://britannica.com/warondrugs)

There is no doubt in my mind that this Anti-Drugg Act Abuse (circa 1986), has messed up our community. Something that we are still suffering to this day in 2022 post 2023. "Since approximately 80% of crack users were African-American, mandatory minimums led to an unequal increase of incarceration rates for nonviolent Black drug offenders." (https://britannica.com/warondrugs)

What can we as Black men do, to cease the attack on us and future generations?

1. We must study and learn our history.

2. We must study laws and the Constitution.

3. Teach what we learn to our children.

4. Support and unify brothers.

5. Realize that we are not our own enemies.

6. Respect ourselves and our community.

"In African American History- slavery, Jim Crow, and the emergence of the northern ghetto- the racial inequality produced by mass incarceration has been perpetuated by the levers of law and political control" (Travis, Jeremey &Western, Bruce:pg295: Policing the Black Man)

SUMMARY

We as Black men must look at how the system has been designed against us. With failure as a nation and as men as the result. From slavery, we can see how our ancestors were in dungeon houses and shipped off the way we are when it is time to hit the penitentiary. Slavery is embedded into our DNA, which is the first step into the western criminal justice system.

When our ancestors thought they were free, they did not realize that the second phase would be Jim Crow laws. These laws and policies continued the first phase of the system. Mass Incarceration is the final blow to the Black community since its arrival in 1973 and the upgraded version in 1982. The Black man has become public enemy number one, these laws are nothing but hindering blocks for the Black man and his family. Today more Black men are labeled as felons, since he has that on his background, he really cannot support his family and set up nothing on his own.

My advice to my brothers is to believe in yourself, and your community. To me this is a major solution that Black men can come and stick together no matter what the circumstances are.

CHAPTER THREE

Can your freedom of speech be used against you?

Can your freedom of speech be used against you?

Chavis Tp hsbAhaw McCray

This essay is dedicated to the youth who aspire to be rappers and choose to rap about street life, trapping, gang banging, hitting links, and so on. Understand that you may want to probably be extremely wise about the words you record and the things you actually do. Watch the things you post on social media. Also, watch who you say it to because people record everything for receipts in 2022. They are screen shotting conversations in chats, text messages, comments, and direct messages and using your own words against you.

It could be a girl or guy you don't like or the feds and other law enforcement agencies would put you in a bind where you trick yourself out of your freedom by incriminating yourself. Earlier this year this was a hot topic of discussion because rapper Young Thug was locked up and charged with a Rico. If you are unfamiliar with what RICO is justia.com states: "The federal Racketeering Influenced and Corrupt Organizations (RICO) law was passed in 1970 as the "ultimate hit man" in mob prosecutions. Prior to RICO, prosecutors could only try mob-related crimes individually. Since different mobsters perpetrated each crime, the government could only prosecute individual criminals instead of shutting down an entire criminal organization. Today, the government rarely uses RICO against the Mafia. Instead, because the law is so broad, both governmental and civil parties use it against all sorts of enterprises, both legal and illegal." (Justia, 2022)

So Thug got the treatment of mafia mobsters because they accuse him of being essentially the ring leader of a Blood street gang in ATL who went by the name of YSL. They hit him with being in a gang and basically committing a bunch of organized crime type of charges. According to Ajc.com he received additional charges of "participation in criminal street gang activity, violation of the Georgia controlled substances act, possession of a firearm during the commission of a felony and possession of a machine gun and drug charges."(Papp, 2022)

How did Thug get here? What are they trying to use against him? Well he got here through being a successful rap artist who sometimes raps some questionable lyrics and says somethings on social media that basically is allegedly being used against him. They locked him and his gang up and put Thug in a position where he has to spend 10s of 1000s of dollars to pay for legal representation while also dealing with his accounts being frozen and people snitching him out to get less time and deflect accountability. He the man with the money so they paint a picture of him as boss who can put money on people heads and make people die. They are trying to tie him to a murder of an individual whom killers are allegedly associated if not apart of YSL that contact Young Thug after the murder.

We don't need see any of our youth go down that road again. So to make an effort to combat a young intelligent brother or sister from headed down any roads similar to Thug's I took the time out to put a presentation I did on the topic in book form to make sure someone smart can avoid this happening to them or can inform someone else. In the words of Great Sweet James Jones aka Pimp C " Ayy, man I just look like this man, knahmtalkinbout? I ain't get this far being no square, man You wanna hide somethin' from black folks, they say you can put it in a book

I don't believe that 'Cause I done read fo' libraries worth of books I got some knowledge y'all need to get up on, man" (UGK, 2007)

In the name of the Pimp I present this game hid in book form to help keep your freedom while not letting it get used it against you.

These are the research questions I laid out to

extract this game I need the reader to soak up like a sponge: 1. Can your freedom of speech be used against you?

2. Can rap lyrics be used against you in court?

3. How are they used?

4. What metric is used to determine whether lyrics are admissible?

5. Do prejudice and bias exist in regards to rap and its lyrics vs another genre?

6 What impact and implications do prejudice and bias on jury and judges present?

Before we can discuss freedom of speech we must be aware of where this right can be found. Here is where it is important to be familiar with the 1st amendment. " Congress shall make no law respecting an establishment of religion, or prohibiting the free exercise thereof; or abridging the freedom of speech, or of the press; or the right of the people peaceably to assemble, and to petition the Government for a redress of grievances."

(https://constitution.congress.gov/constitution/amendment-1/) This amendment comes from our constitution and to summarize it in the context of this essay no law can be made to keep you from exercising your freedom of speech. Interestingly, if you have ever seen someone get arrested one of the first things law enforcement MUST do is read your Miranda rights. "In the United States, the Miranda warning is a type of notification customarily given by police to criminal suspects in police custody (or in a custodial interrogation) advising them of their right to silence and, in effect, protection from self-incrimination; that is, their right to refuse to answer questions or provide information to law enforcement or other officials.(Miranda v. Arizona". Oyez. Archived from the original on September 5, 2019. Retrieved September 23, 2019.) To a second and critically think to yourself if you have the right to free speech why would a police officer have inform you of your right to remain SILENT..? If you were paying attention to what you just read the part about "protection from self incrimination" should have made a light bulb go off. There is essentially a catch 22 where you can say whatever you want but at the same time it can be used against you. You are held accountable for what you say and do though you can do and say what you please. Some extra game is if you aren't read your Miranda rights you can have a case thrown out against you. To show that I'm not making this up I present a quote from Thiessan law firm which states " If you were not read the Miranda Warning before being questioned and there is not a valid waiver of these rights, any statements or confessions you made during the interrogation are deemed as "involuntary" and cannot be used against you."(Thiessan, 2022) This information should sufficiently demonstrate why the answer to question 1 is yes and help you navigate your way through combating it if you are ever put in this position. To summarize,

First amendment provides you with the right to pretty much say what you want freely while at the same time the Miranda warning provides you with a right of silence protecting you from incriminating yourself. The paradoxical window created here opens up an opportunity for your freedom of speech to be used against you.

To answer the 2nd question of whether rap lyrics can be used in court I present this excerpt: "Courts in several states have allowed prosecutors to introduce rap lyrics written by defendants into criminal trials as evidence of motive and intent. The American Civil Liberties Union ("ACLU") determined that courts admitted defendants' rap lyrics into evidence at trial in almost 80% of cases examined from 2006 to 2013. "(Brief of Amicus Curiae ACLU of New Jersey, New Jersey v. Skinner, 95 A.3d 236 (N.J. Sup. Ct. 2014). So not only can they but they are being used at a disturbingly high rate.

The 3rd question of How this evidence is used is described here via explanation of exactly what rap lyrics are evidence of specifically. ""In general, evidence of a criminal defendant's past bad actions cannot be introduced to show that he or she had a bad character and acted consistently with that bad character on a particular occasion (such as when the alleged crime occurred). This "character evidence" is acknowledged by the rules of evidence (such as Rule 404 of the Federal Rules) as having a unique tendency to prejudice jurors against the defendant. But, evidence that falls into the category of character evidence may be admitted for purposes other than to show that the defendant acted in accordance with the character trait."(https://www.criminaldefenselawyer.com/resources/rap-lyrics-evidence-is-it-a-crime-rhyme.htm) The character evidence rules come from FRE 404, which prohibits the use of evidence of a defendant's character to prove that they had the propensity to commit the charged crime.

The rule further provides exceptions to this rule, including FRE 404(b)(2), which allows character evidence to be used in a criminal case to prove "motive, opportunity, intent, preparation, plan, knowledge, identity, absence of mistake, or lack of accident." Prosecutors aiming to use a defendant's rap lyrics against him or her typically utilize the exception granted under 404(b)(2) to argue that the lyrics demonstrate prior knowledge or intent to commit the charged crime.Judges are supposed to weigh the probative value of this evidence against its undue prejudice to the jury before deciding whether to admit.

To answer the 4th question FRE 404 is essentially the metric admissibility is gauged by. "Congress and the drafters of the FRE recognized the need for a narrow exception for non-propensity character evidence, which they codified in Rule 404(b)(2). Ultimately, this exception became the most litigated rule of evidence in the FRE. The current text of Rule 404(b) states that: "Evidence of a crime, wrong, or other act is not admissible to prove a person's character in order to show that on a particular occasion the person acted in accordance with the character . . . This evidence may be admissible for another purpose, such as proving motive, opportunity, intent, preparation, plan, knowledge, identity, absence of mistake, or lack of accident"

Fed. R. Evid. 404(b) (2021) An inclusionary and exclusionary approaches are used the article adds "Wigmore's exclusionary approach assumes exclusion of evidence of prior "crimes, wrongs, or other acts" unless it is offered only to prove "motive, opportunity, intent, preparation, plan, knowledge, identity, absence of mistake, or lack of accident." In contrast, Stone's inclusionary approach is a broader interpretation that generally allows evidence of a defendant's prior crimes, wrongs, or other acts unless it is offered solely for the purpose of showing the defendant's propensity to commit the crime in question. The key difference is that the inclusionary approach broadly permits propensity evidence under Rule 404(b)(2), while the exclusionary approach only allows evidence of prior bad acts if it is used for a non-propensity purpose."(Sripathi. V, 2022)

Question number 5's answer can be directly answered with this excerpt: " Dr. Fischoff found that the criminal defendant was viewed as more likely to have committed the murder when the jury knew he authored the rap lyrics than if they did not know he authored the rap lyrics. Moreover, the study showed that mere authorship of the rap lyrics was "more damning in terms of adjudged personality characteristics than was the actual fact of being charged with murder." These findings demonstrate a substantial jury bias against rapper-defendants and suggest that extant negative stigmas surrounding rap music are exacerbated when rap lyrics are admitted into trial."(Sripathi. V, 2022) This is objectively not fair when we understand that majority of the time it is the African American who aspires to rap, gets racially profiled more by police, and eventually get locked up at rates higher than others through America's historically oppressive and racist judicial system in regards to black people. Not only can we confirm prejudice and bias exist we can demonstrate that this unfair outlook and reasoning isn't applied in other genres. "Sociologist and criminology professor Charis Kubrin, who has consulted on forty cases as an expert witness on rap music, recognizes that the literal interpretation of rap lyrics itself poses a "horribly prejudicial" effect on juries. In a study she conducted at the University of California, Irvine, Kubrin found that subjects who believed they were reading rap lyrics interpreted them "very literally" as opposed to subjects who believed they were reading country lyrics—even though the lyrics were the same."(Sripathi. V, 2022)

The last question can be answered in the same article when it states "A 2017 study conducted by Adam Dunbar (and supervised by Charis Kubrin) built upon prior research to examine the impact of rap lyrics on jury perceptions. In a series of experiments, this study found that "participants deemed identical lyrics more literal, offensive, and in greater need of regulation" when the lyrics were identified as rap music as opposed to country. Further, the study revealed that the songwriter of the lyrics was viewed "more negatively across a number of dimensions when the lyrics were categorized as rap" rather than other genres. Most damning, however, was the finding that subjects suspecting a defendant to be guilty were more likely to treat rap lyrics as an "admission of guilt" in the trial context."(Sripathi. V, 2022) A defendant shot at beating this type of case with cards stacked against them in this way is very slim. We are supposed to be innocent until proven guilty and these prejudices and bias disable that fair process.

Some states recognize this unfair dynamic and have attempted to find a solution. "The New York State Senate passed a bill yesterday (May 17) limiting the use of song lyrics as evidence in court by prosecutors.

First touted last November, the purpose of this bill is to set a new high bar compelling prosecutors to show "clear and convincing evidence" that a defendant's rap song, video, or other "creative expression" is "literal, rather than figurative or fictional".Led by Brad Hoylman (D-Manhattan) and Jamaal Bailey (D-The Bronx), the bill – called 'Rap Music On Trial' and dubbed Senate Bill S7527 – has received support since the idea was raised from Jay-Z, Run The Jewels' Killer Mike, Meek Mill and more."(Richard,2022) Though that seems like relief states like Georgia aren't letting up. Georgia DA Fani Willis was quoted saying "I think if you decide to admit your crimes over a beat, I'm going to use it," Willis told reporters at the news conference. "I'm going to continue to do that; people can continue to be angry about it. I have some legal advice: Don't confess to crimes on rap lyrics if you do not want them used — or at least get out of my county."(Bellamy-Walker, 2022)

In Conclusion, yes your freedom of speech can be used against you. Rap lyrics are an expression of this freedom that can be used against you to demonstrate character evidence, in certain courts demonstrate your propensity to commit a crime, and also any other demonstration of you planning, being aware of or any other forms of self incrimination. FRE 404 is essentially the metric admissibility is gauged by. Bias and prejudice does exist according to multiple studies examining these sociological phenomenon that occur in regards rap music when comparing and contrasting with other genres. The implications and impact are heavily associated with being applied in charging and conviction of African Americans who participate in rap versus an almost non existent issue in regards to genres like rock, country, and others with questionable lyrics that rarely get seen as anything other than an artistic expression that wasn't meant to be interpreted as literal. Where as rap lyrics and artist don't get the same benefit of doubt.

CHAPTER FOUR

Towards a deeper understanding of Indigenous African Information Systems as a catalyst for ideological, socio-economic, and holistic liberation.

Towards a deeper understanding of Indigenous African Information Systems as a catalyst for ideological, socio-economic change

Robert M.

From an early age I knew I wanted to be a scientist of some sort. I was always interested in the observation and understanding of the natural world. I also had a strong desire to see people who looked like me demonstrating competency and mastery of their intellectual domain. In my reading I would encounter historical figures like Benjamin Bannekar or George Washington Carver. It was always fascinating to see men of African descent who excelled in fields of science and technology despite being born into a construct (blackness) created by a racially oppressive society. It definitely gave me a sense of pride to be able to read about men of distinction who looked like me, but I still felt a deep sense of internal conflict. I was not in almost any sense satisfied with the picture of Africa that had been presented through the western lens. In particular, the westernized image of Africa and Africans below the Sahara that portrays the land and the people as being almost completely devoid of value. My dissatisfaction with this portrayal of Africa led me on a journey that was fueled by an intense curiosity, a deep love for reading, science, technology, and a genuine desire to understand my own identity as a member of the African diaspora.

Virtually from the very beginning I had two somewhat seemingly conflicting concepts in my mind, a genuine inborn love of science and technology but also a desire to have a deep spiritual connection to what I viewed as my roots in the African experience. I learned fairly quickly that the western paradigm was geared towards painting a "less than human" picture of Africa. On the other hand, my innate love of science was so deep that the western bias towards Africa could never truly persuade me to totally accept the persistent narratives. In the words of René Descartes "Cogito, ergo sum", "I think therefore I am". As a thinking individual, I could never accept the idea that the people who looked like me were incapable of producing anything remarkable, especially in the realm of science, art, and society. I recall reading the words of Thomas Jefferson who voiced his opinion about the descendants of Africans when he said that he felt the black man was much more inferior in reasoning to the white man and incapable of pondering upon abstract ideas. I also made note of the letter that Thomas Jefferson received from one Benjamin Banneker. Benjamin Bannekar was a brilliant mathematician and surveyor, a free American man of African descent in the late 1700s. Among other things, he authored his own Almanac and built his own mechanical clocks. Here Banneker, a man who proudly claimed to be of African descent, his very existence challenged the popular ideas of the time held by many whites towards people like him. Banneker also challenged Thomas Jefferson in a way that only a deeply thinking individual could when he pointed out the hypocrisy of slavery and the surrounding propaganda that down-played the capailities of African descendants (and by proxy Africans), while at the same time embarking upon a unique experiment in a democracy that supposedly placed human rights as the lynch pin of the principles that would guide the society. It's hard to imagine that Banneker's words didn't hit Thomas Jefferson square in the chest. Banneker demonstrated that as a free black man he understood very well the concepts that Jefferson laid out in the Declaration of Independence. In fact Bannekar challenged

Jefferson by pointing to the core idea of the Declaration of Independence, that all men were born with innate rights that no other man had the right to violate. Banneker's knowledge presented Jefferson with what had to be paradoxical to his own worldview. If this black man (Banneker) could not only understand but also articulate the deep philosophical concepts of human freedom that Thomas Jefferson himself learned from European philosophers like John Locke, then his previous general assessment about the capacity of Africans to ponder upon abstract ideas was invalid. It's interesting that Jefferson seemed eager to respond to the letter. In just two weeks Jefferson responded to Banneker's August 19th 1791 letter with his own letter written on August 30th 1791. Jefferson praised Banneker's skill and accomplishments, calling him a "credit to his race". Jefferson also reversed the previously held opinion about blacks being intellectually inferior to whites. While it's interesting that Jefferson never touched on the issue of slavery in his response to Banneker, I find it even more interesting that Jefferson yet again reversed his opinion about blacks mental facilities just a few years later. This ideological exchange between Jefferson and Banneker, more importantly, Jefferson's own internal conflict, highlights the intrinsic difficulty of extracting even objective views about Africa and Africans from the western world. If a supposedly "enlightened" white man, an Anglo-Saxon in particular, could not commit to at least establishing that African descendants deserved the right to individual freedom, how much less will others even care to question if the humanity of African descendants is something worth considering?

This western bias towards Africans presents the backdrop for the dilemma that Africans and African descendants face. How can we move towards a renaissance of Traditional Indigenous African Information Systems or even begin to reconstruct and reimagine the potential value to the modern western context without addressing the internal conflict caused by our own westernization. To move towards a deeper understanding of African Information Systems, we have to reconcile these two competing halves of ourselves. In his book "The Souls of Black Folks", W.E.B. Dubois calls these two halves of African descendants a "double-consciousness", the idea that we (African Americans) look at ourselves through the lens of the western world, and that we measure ourselves against western metrics, standards by which we will almost always seem to fall short of. Dubios observed astutely, "The history of the American Negro is the history of this strife,—this longing to attain self-conscious manhood, to merge his double self into a better and truer self. In this merging he wishes neither of the older selves to be lost. He would not Africanize America, for America has too much to teach the world and Africa. He would not bleach his Negro soul in a flood of white Americanism, for he knows that Negro blood has a message for the world." It's ironic that while a very disproportionate number of Africans and African descendants join in with the western chorus to devalue Africa, the western world is heavily capitalizing materially off not only Africa's natural resources but also knowledge that was gleaned from Traditional Indigenous African Information Systems. Imagine if we African descendants (at home and in the diaspora) saw the same value in Traditional Indigenous African Information Systems and invested as much intellectual energy into understanding them as the West invests in exploiting and capitalizing on African geological, biological, and ideological exports. Imagine if we tuned our westernized minds into the frequency of the African blood that cries out in our veins, however with a renewed sense of scientific and philosophical discovery. I believe that not only can we regain that aspect of ourselves that was lost, but we can also unlock the message that Traditional African Informations Systems has for the world.

In my journey into manhood, I was inspired by many things. I was inspired by music, by books, by technology, by the culinary arts. The source of these inspirations were my parents, and even though the information was often transmitted verbally, there was a tremendous amount of information that was also transmitted through symbiotic observation. My brain recorded my father playing classical R&B music on his 8-track or writing computer programs on his Radio Shack TRS-80. My brain recorded my mother being a creative and expressing herself through the love of art, interior decoration, and cooking. What manifested through my observation of my parents were the building blocks that would influence me to become a musician, a computer scientist, an intellectual, and a pretty decent cook. What I did not know then was that what I experienced as a child was basically the fundamental template of African information systems. Having increased my understanding of technology as a technologist, and building deeper connections to Africa through learning, I began to see a unique opportunity for people of the African diaspora to leverage their "double-consciousness" as westerners and as people of African descent to construct a new identity. This journey of merging our two selves, can be facilitated by relearning and re-examining traditional African information systems while adding to it the application of western concepts like academic rigor and the scientific method. Merging these two aspects of ourselves, we will be able to build new pathways that will allow us to revitalize our communities and create opportunities to improve the collective status of our people holistically (physically, psychologically and socio-economically).

Many of us have navigated our way through the early years of the collective effort to revive an African centered worldview from the ashes of slavery and oppression. The early effort towards resurrecting an African worldview can be compared to fishing the depths of the Atlantic for the bodies, minds, and voices of long gone ancestors. Much of the early information that inspired the consciousness movement was highly outdated or purely pseudo, relying on often oft repeated myths or ego driven ideologies to boost the self confidence of African descendants. However, now that many of us are beginning to value the type of rigor needed for academic research and the application of the scientific method, we should invest in the collective intellectual effort to build cultural and ideological bridges back to traditional African information Systems. With this renewed perspective, we can incorporate certain elements that we learned from western society, like observation (experimentation and documentation) but also heed the lessons we are learning in this society as we grapple with corporate greed and the dangers of unchecked capitalism.

Now that we are decidedly in the process of course correction, we can begin to feed our curiosity by observing the groundwork for a fresh approach to Africa that has already been laid by quite a few African scholars. The process of re-examining, reclaiming, and reinterpreting the history of the indigenous African worldview has been underway in the interior. The torch has been lit and we should prepare ourselves to be strong enough, swift enough and adept enough to carry the baton for our leg of the diasporic African race into the future.

According to Frances E. Owusu-Ansah and Gubela Mji "African knowledge is experiential knowledge based on a worldview and a culture that is basically relational. The spirit of the African worldview includes wholeness, community and harmony which are deeply embedded in cultural values." (Frances E. Owusu-Ansah and Gubela Mji). Because we have westernized cultural values, our journey back into understanding the African world view has been somewhat convoluted by the colonized mind. As a technologist, I bounced this idea around in my head about how I could look at Africa from the context of technology and innovation. It quickly occurred to me that my intellectual energy was already taking me down the wrong path before I even began putting pen to paper. What I really wanted to do was to resist falling victim to low hanging fruit and getting into this mode of trying to build up some fantastic narrative that would be "hefty" enough to compete with the heavy hitters of classical western European societies. I doubt any intellectually honest person would come to the conclusion that there is no merit in westernized information systems. However, I think many would agree that the Westernized world view is in conflict with the world view of Indigenous cultures in many respects. The westernized world view sees humanity as the pinnacle of the known universe, while the indigenous world view sees humanity as not only a protector but also an active participant in the circle of nature. One example that highlights the exploitative nature of the western world view is a report by the Intergovernmental Science-Policy Platform on Biodiversity and Ecosystem Services (IPBES). According to IPBES a quarter of our planet is managed (owned, used, and occupied) by Indigenous populations. On average, indigenous populations are doing a better job of managing natural resources and environmental hazards like species decline and pollution. Considering that Africa contains 90% of the planet's biodiversity, the stakes are high for the need to reintegrate traditional African knowledge systems back into the cultural consciousness of indigenous Africans and their relatives throughout the diaspora.

African Ways of Knowing (Traditional African Knowledge Systems)

Traditional indigenous African education is a holistic way of learning that has been used since the beginning of time to pass on information and values from one generation to the next. This type of education is rooted in the ancient culture, spirituality, and traditions of the African people. It is based on values like respect for elders, respect for nature and the environment, and an understanding of one's place in the community. The purpose of traditional education is to teach young people about their culture, traditions, and history, and to equip them with the knowledge and skills needed to be successful in life.

Just like now in the computer age, the way that humans share information has been a vital aspect of society. Examining Traditional Indigenous African information systems, The African world view is experiential and oral in as much as the Western World view is theoretical, empirical, and document heavy. Traditional Indigenous African education systems naturally emphasize immersion. Traditional African Information Systems have played a vital role in African societies for centuries. Their primary purpose is to record and store important information related to agricultural practices, trade, legal matters, social rituals, and much more. African Information Systems are typically passed down orally, making them particularly sensitive to oral tradition and story-telling. These systems involve collective remembering and involve regular constructions and reconstructions of cultural knowledge. Furthermore, these systems often store information in poetic verses, songs, and other means. Overall, traditional African Information Systems are powerful tools for storing, sharing, and preserving knowledge within African societies.

Modes of Learning

Studies on traditional African hunter gatherer communities have shown that the primary learning modes observed among them are vertical (from parent to child). Among Aka hunter–gatherers, about 80 per cent of their knowledge about subsistence, childcare, sharing and other skills were transmitted from their parents, generally from the same sex parent (Hewlet et al). Vertical learning is a form of learning that moves from the basics of a subject, to more challenging and advanced concepts. This type of learning is highly valuable because it allows students to develop a strong foundational understanding, before exploring the nuanced details of a subject. As students move up the "ladder" of knowledge, they are better prepared to tackle complex topics and graduating levels of difficulty. This makes vertical learning an effective way to gain a deep and thorough understanding, while preparing for higher levels of proficiency and comprehension. African Pygmies are among the most remarkable and unique societies of the world. Their traditional society is an example of vertical learning, a process in which knowledge and skills are passed down from elder groups to younger generations in a way that preserved the stability and wellbeing of their culture for thousands of years. This type of learning has been fundamental in allowing the Pygmies to become an integrated part of their regional ecosystems by understanding specific plants and their uses, learning to understand and interpret animal behavior, distinguishing between safe and dangerous game, fishing techniques, and a variety of other skills necessary for their basic survival. The Pygmies' vertical learning process is almost entirely exclusively oral. While this type of learning is not heavily documented, it is likely that experience and knowledge passed down within the Pygmy family lines are accompanied by traditions and rituals. Through these traditional methods, the Pygmies have been able to survive for years without any outside knowledge or interventions. Another intriguing aspect of social learning modes among the Pygmies is the observation by sociologists of a very high self-motivation that seems to happen very earlier and more frequently for Pygmy children than other human societies geared towards different means of sustenance and production (It would be interesting to conduct deeper research on the connection between the forager lifestyle and accelerated learning in young children). Because of the importance of information and knowledge passed down from older to younger generations, the Pygmies have seen a decrease in their cultural practices in recent years due to societal changes. Development projects, deforestation, and other activities have interfered with the traditional Pygmy lifestyle. In order to restore vertical learning and to ensure that the Pygmies can continue to benefit from the knowledge and skills of their elders, it is important to keep their culture alive. This includes supporting projects that connect elders with younger generations, such as traditional craft-making and storytelling, in order to ensure that their unique knowledge is not lost. Also, another unique opportunity is presented here in that in the process of helping Pygmies reconnect, modern information systems can be used to document the valuable knowledge that is contained in their societies about plants, animals, and the items traditionally used for medical purposes.

Traditional African Ways of Healing

One of the areas where there is a stark contrast between the western world view and traditional African information systems is in medicine/health care. In African Traditional Medicine (ATM), "Healing is a part of the complex religious attempt by Africans to bring the spiritual and physical aspects of the universe as well as man who lives in it, into a harmonious unity and wholeness. Wholeness is therefore the underlying focus of African medicine." (Onah Gregory Ajima* and Eyong Usang Ubana). African medicine does not view healing as an effort aimed strictly at addressing physical ailments, it views sickness or disease as being out of alignment with nature or the universe. African Traditional Medicine acknowledges both the physical and metaphysical aspects as a pathway to healing (wholeness). In recent years we've seen Western medicine "catch up" in the sense of considering positive attitudes and spirituality as playing an important role in health outcomes. According to a study led by researchers at Harvard T.H. Chan School of Public Health and Brigham and Women's Hospital, "spirituality in serious illness and in health should be a vital part of future whole person-centered care, and the results should stimulate more national discussion and progress on how spirituality can be incorporated into this type of value-sensitive care."

African Traditional Medicine (ATM) is a type of health care system that relies on centuries-old healing methods that employ indigenous plants and herbs as remedies. ATM is practiced in many places in Africa and is considered to be one of the important aspects of traditional African culture. This type of health care has been passed down from generation to generation and is used widely across Africa by both individuals and in medical facilities.

The use of plants and herbs as medicine is one of the primary aspects of ATM. These remedies, which are often concocted into teas, are thought to offer relief from a wide range of physical and psychological ailments. Additionally, some traditional remedies are based on traditional beliefs, such as potions and charms meant to ward off evil spirits.

In recent years, scientific study has been done to explore the efficacy of ATM. In many cases, herbal remedies have proved to be effective in treating various conditions such as headaches, fevers, and stomach upsets. Despite this, some skeptics remain unconvinced of the effectiveness of traditional medicines. Regardless, many in Africa remain committed to their traditional practices and beliefs and attitudes towards ATM are starting to shift. In many places, ATM is accepted by the mainstream medical establishment and practitioners are working together to explore the potential of traditional health care practices.

The irony regarding the West's "skepticism" in ATM is highlighted by the fact that Western pharmaceutical companies have long profited from African Traditional Medicine (ATM). Large companies such as Pfizer and Merck have made significant investments into local traditional medicine, seeking to produce and market treatments derived from African herbs and plants. These treatments, often referred to as 'Westernized' versions of traditional medicine, have become extremely popular in the West and have provided these companies with millions of dollars in profits. In addition to producing their own treatments derived from African herbs, many western companies have also bought rights to the traditional healing knowledge of African peoples. This has allowed them to patent African medicines, allowing them to make even more money from the sale of these treatments.

As more western pharmaceutical companies invest in the production and sale of traditional African medicines, the profits have continued to skyrocket.

Despite the financial successes of western companies, African peoples have gained almost no benefit from their traditional medicines being exploited. In many cases, profits from the sale of these treatments go to large corporations and their shareholders, leaving the traditional healers and the communities that rely on their treatments with very little in return. This has further entrenched poverty in many African communities and has made it increasingly difficult for those living in these areas to receive adequate healthcare.

The Opportunity, The Promise, and The Risk

I find it very interesting that out of hundreds of years of Western European Imperialism and propaganda against Africa, we have been deliberately kept in the dark about the tremendous wealth in natural resources (including humans) and the significant looting of that wealth on every level. Even with the rise of globalism and the decline of European Imperialism Africa is still very much being robbed of precious resources with very little benefit to Africans. At this moment in time there is an opportunity for Africans and people in the Diaspora to help Africa see her value to the world and reclaim the right to manage the distribution of her resources.

The promise I believe is that even with the tremendous amount of looting of Africa's resources that has already taken place, there is still so much more that she can give to the world. I would hope that Africa's children (essentially the whole world, but more significantly her children that were looted in the process of the transatlantic trade) would recognize their stake in Africa, and also the responsibility we have to reflect a mirror in the face of humanity and demonstrate how far we've moved from lifestyles and traditions that were built on concepts that were focused on being in alignment with nature and the world around us (stewards or conservators of the earth) to a lifestyle of becoming indifferent to over-consumption to the point that we show virtually no concern for the generations that will follow.

The west has already begun the process of codifying and documenting the practical aspects of African Information Systems. The west has been studying things like traditional African learning modes, and documenting medicinal herbs and plants based on indigenous African knowledge. The one significant area where the West has been stifled by African Information Systems is in the fact that traditionally African societies shroud the rituals and concepts related to the metaphysical aspects of their information systems in secrecy. Typically these "secrets" in Traditional African Information Systems have been a barrier to western observation because they are "protected" by a select group who are initiated and sworn to protect the secrets. However, if the keepers of these "secrets" become extinct then that knowledge is potentially lost for generations or perhaps forever lost to time. Secret societies have been a part of the cultural heritage of sub-Saharan Africa for centuries. These societies often served as initiatory and religious organizations, with their activities shrouded in secrecy and mystery. Membership was restricted to certain classes or tribes, and often involved rigorous initiation ceremonies, rituals, and oaths of secrecy. Some of the most well-known secret societies in sub-Saharan

Africa include the Ekpe of the Efik people of Nigeria and the Poro of the Mende people of Sierra Leone. These societies played a significant role in the social, political, and economic life of their communities, and were often involved in the resolution of disputes, the maintenance of law and order, and the promotion of social cohesion. Who would stand a better chance to become an initiate or a protector of the knowledge and information contained in these societies than children of the Diaspora?

We now have the knowledge and the technical understanding that was ushered in by the rise of the Western World, followed by globalization. We can now marry the Western perspective with the concepts embodied in the Indigenous perspective to bring about potentially transformative change. Maybe what we have to teach to the rest of the world can be unlocked in the process of becoming reconnected or grounded in certain elements of traditional African information systems. I believe that as descendants of the diaspora but also as children of the West, we have a tremendous opportunity to make ourselves "whole" and attain a higher degree of conscious personhood. The risk is that we have fallen so in love with western ways that we are not willing to embrace perspectives from the ancestors that might appear to be counter intuitive or antithetical to western perspectives.

CHAPTER FIVE

Assessing Emotional Intelligence and its Significance to African Americans

Assessing Emotional Intelligence and its Significance to African Americans

Chavis Tp hsbAhaw McCray

What inspired this topic and how?

This topic was inspired by a Facebook post of a fellow honorary tep heseb member and creator of the Facebook group, which the title of this publication is named after. My good brother Mike Rainey has always been a very knowledgeable person whom I interacted with early in the beginning stages of what the black social media community calls consciousness. This is a timeframe in the beginning stages of your awareness and paradigm shift in our journey in this phenomenon we call life. This brother was extremely passionate about truth and knowledge seeking with a particular mutual interest in the history, culture, language, and philosophy of the ancient remetch (Egyptians). We both resonated with the powerfully conceptualized idea of Maat, and I was heavily into the language and learning to decipher script and debunking misconceptions of black Hebrew Israelites, Christians, Muslims, and self-proclaimed Moors in online Facebook groups. He recognized my drive and intrinsic motivation to edify my peers with accurate information and understanding, so he asked me to be an admin in his Facebook group. We were tired of false information on the subject and misinformation in general, so we went on a mission to counter nonsense with quality information and build our platform to be the change we wanted to see and do something bigger than us creating an online community I'm which we could collectively network and flood social media with the dissemination of "right knowledge" and understanding arming our loved ones and other truth and knowledge seekers with an archive of well-researched information and credible, viable resources to on their own time edify themselves and others. We both loved to learn and shared information back and forth so much that he one day purchased himself and me a 50$+ book with more than 500 pgs that I have still yet to finish on what I call the G (for free without looking for anything in return but to study with him and keep sharing information). I one day randomly saw him mention the topic and immediately told him he should write about it in this book I wanted him to participate in with me and others. After looking into the topic on my own, I felt an obligation to share what I learned from that random moment in time and put it in this book myself as only I could touch on it in the way I felt necessary, so I am doing it myself.

What is emotional intelligence?

The topic of concern is the concept called emotional intelligence. "Emotional intelligence refers to the ability to identify and manage one's own emotions, as well as the emotions of others. Emotional intelligence is generally said to include a few skills: namely emotional awareness, or the ability to identify and name one's own emotions; the ability to harness those emotions and apply them to tasks like thinking and problem-solving; and the ability to manage emotions, which includes both regulating one's own emotions when necessary and helping others to do the same." (https://www.psychologytoday.com/us/basics/emotional-intelligence). Being that this publication has the youth in mind, the things the concept refers to would benefit one to make a part of the foundational collective consciousness to apply where necessary while maneuvering through life and the adversity that reveals itself on a day-to-day basis.

Why is emotional intelligence important?

Emotional intelligence is important because it is essentially a metric to gauge your ability to self-reflect via your self-awareness to, at times, read the room or possibly control how the room is read. Being book-smart is just one form of intelligence. Self-awareness and the ability to apply critical thinking and problem-solving skills to actual social interactions, added to book smarts, are essentially the building blocks needed to breed leaders. Raising a family first and community second, well-rounded social and academic problem-solving community member. To further elaborate on why an emphasis on emotional intelligence is of significance, psychology today states: " A person high in EQ is not impulsive or hasty with their actions. They think before they do. This translates into steady emotion regulation, or the ability to reduce how intense an emotion feels. Taking anger or anxiety down a notch is called down-regulation. The emotionally intelligent can shift gears and lighten the mood, both internally and externally. Such people are especially tuned into the emotions that others experience. Understandably, sensitivity to emotional signals both from within oneself and from one's social environment could make one a better friend, parent, leader, or romantic partner. Being in tune with others is less work for others. This person can recognize and understand the emotions of others, a skill tied to empathy. A person with a high EQ can hear and understand another person's point of view clearly. The empathic are generally supportive of the people in their lives, and they easily modulate their emotions to match the mood of another person as well."(PsychologyToday, 2023) Imagine if there was an entire ethnic group that collectively had a paradigm shift in the direction of focusing on emotional intelligence just as much as intellectual intelligence. Think about the community we live in. How rampant is impulsive behavior? How many of our people's freedom and lives have been lost forever behind someone's hasty actions? The logical counter to this would be to engrain the opposite type of behavior and mindset because it is the root cause of the issue at hand. But this is easier said than done.

How does culture influence emotional intelligence?

Part of the reason I say It is easier said than done is that it isn't quite that simple. There are dynamics at play that are perpetually countering your counter. While researching this topic and doing a literature review on several academic resources on the related sources, I came across an article that I feel brilliantly addresses the question of "How does culture influence emotional intelligence?" I felt it deserved to be in a book. Titled Emotional Intelligence In African Americans: How Culture Affects The Applicability of the Construct.

Kristen Y. Sanders, M.A. in 2021 states: There is an abundance of research (e.g., Kwon et al., 2013; Qu & Telzer, 2017; Ma et al., 2018; Nozaki, 2018; Tamir et al., 2015) that studies the direct effect that culture has on an individual's emotional expressiveness. Culture has been found to influence the beliefs and values that individuals hold concerning emotions and how they should address them. One common area of research is the difference between East Asian and Western cultures in their styles of emotional expression and emotion regulation. It is a common assumption that Western cultures value individualism while Eastern cultures value collectivism. Individualistic cultures promote autonomy. Emotions are viewed as important components of the self and should be expressed as such, with integrity. In collectivistic cultures, autonomy is not as important. The family or community functions as one and individual decisions are made with others in mind (Qu & Telzer, 2017). Western cultures view emotion regulation as less important. Emotions are viewed as assets to one's experience and are expressed openly (Qu & Telzer, 2017). Westernized individuals pride themselves on being honest and true to themselves. They may not feel it is necessary to regulate what they are feeling in response to certain situations. Because of this, they may view the self and personality as stagnant, or fixed across situations. There is less of a need for regulating emotions because they are viewed as acceptable and unique to each individual (Qu & Telzer, 2017). Emotional expression in African Americans is related to their experiences within society (Nelson et al., 2012). Boykin, as cited in Nelson et al. (2012), proposed a Triple Quandary Theory that identifies African American culture as one that "emphasizes" the importance of emotional expression, oral communication, and strong communication between children and their parents (p. 2). It is the African American experience, plagued by oppression, that shapes how they relay their emotions to the rest of the world (Consanine & Magai, 2002). Instead of expressing emotions freely, they choose to repress or blunt them to avoid repercussions, resulting in the emphasis being shifted toward limited self-disclosure and emotional self-control. This is said to affect the degree to which certain emotions, especially those that are negative, are felt, expressed, and reported (Consanine & Magai, 2002; Plasky & Lorion, 1984). The societal stressors that African Americans face force them to operate with caution when expressing their emotions due to the fear of being negatively evaluated by others (Consanine & Magai, 2002).As stated, African Americans have a unique experience plagued by societal pressures that affect their life experiences. Their means of coping with these stressors are often viewed as maladaptive and linked to negative mental health outcomes (Graham et al., 2015). These styles of coping and emotional expression styles are deeply rooted in culture, much like other researched groups. From an EQ perspective, African Americans exhibit a style of EQ that is unlike the EQ found in the dominant cultures and thus regarded as a lack of EQ rather than a different style. Many of the studies on the emotional regulation style of African Americans

characterize it as one that is dysfunctional. According to Barbarin (1993), the topic of poverty as it relates to African Americans is often brought up when considering what contributes to said dysfunction. It Is true that for many African Americans, poverty is an issue, and children living in impoverished conditions are at risk for higher levels of developing externalizing behaviors and experiencing social and academic maladjustment. While this is true, it is also found that many African American children are performing exceptionally in their academic abilities, social competencies, emotional health, and family/community life despite the risk (Barbarin, 1993). Barbarin identified resilience and coping as two factors that deserve more attention when studying emotional expression and regulation in African Americans. According to Supplee et al. (2009), emotion regulation is defined as "the ability of an individual to transform an emotion or to devise coping mechanisms to manage emotions" (p.394). It is said to begin its development early in life and continue throughout one's toddler and

preschool years. Supplee et al. looked to identify the link between emotional regulation strategies(ERS) and externalizing behaviors in both African American and European American children. It was noted that culture was assumed to affect parental attitudes toward certain coping behaviors. The coping behaviors accepted by European American parents (e.g., physical comfort-seeking, thumb-sucking) may be frowned upon by African American parents. It was also found that ERS may have differing outcomes in children from different ethnic groups. Strategies that involved methods of self-soothing were linked to low externalizing behaviors in European American children but high externalizing behaviors in African American children. This is said to be dependent upon the culturally rooted attitudes toward the ERS. Certain strategies are promoted by one group because they are more effective for that group. Given that the children's ERS were reported by their parents in the study, if a certain strategy is viewed positively by a parent, then they are more likely to rate their children lower on problematic, externalizing behaviors. African American mothers were more likely to promote more mature coping behaviors with their children (Supplee et al., 2009)

It is well-documented that African Americans are more susceptible to anxiety-related disorders than other groups. This is said to be due to the inundation with racism and discrimination that African Americans may endure. Due to a lack of awareness and resources, African Americans are more at risk for untreated mental disorders, with anxiety being one of the most prevalent (Graham et al., 2015). Hollard, as cited in Graham, Calloway, and Roemer (2015), suggested that dysregulation of emotions is "at the core of anxiety disorders" (p. 554). This would suggest that the reason behind the anxiety that African Americans feel is related to their style of emotion regulation. The findings of the study conducted by Graham et al. supported this suggestion. The "difficulties" in emotional expression were found to be significantly correlated with anxiety symptoms. Individuals who had more difficulties in regulating their emotions also reported high levels of anxiety. The researchers proposed that providing African Americans with different types of coping strategies may alleviate their symptoms (Graham et al., 2015)." (Sanders. K, 2021)

How clear are you now on How culture influences emotional intelligence after that lengthy, clear, informative, and edifying excerpt? Some may complain that I didn't summarize the paper and those that may charge me with plagiarism for citing what I did but be clear the author was cited, and I am in no way attempting to pass off this as something I authored or take credit for. Hence, me properly using in-text APA citations and making sure the author is cited in my references at the end of this piece. Though this isn't groundbreaking research and the information is publicly available, the reality is nine times out of 10, the readers who take an interest in our publications non, readers probably wouldn't do the extensive and meticulous combing through different journals and resources actually reading and comprehending what they read to enough to even be able to weave this brilliant paper and the topic of discussion together and present it in a way that it is by default edifying and I, as the author hope you can see how powerful this understanding you just obtained, was concerning the question that was asked earlier at the beginning of this portion. Nothing more needs to be said.

Is there research on emotional intelligence in the African American community or on emotional intelligence to reduce trauma in Africana American males?

" Research reveals that although there have been several systematic barriers identified that impact mental health in the African American community, there has not been a sufficient amount of research on emotional intelligence in the African American community or on emotional intelligence to reduce trauma in Africana American males" (Brandford. D, 2020). This quote comes from Davis Brandford at Clark University's paper titled "The Mental Health of Black Men: Stabilizing Trauma with Emotional Intelligence. I thought that it was intriguing that this author thought emotional intelligence was so significant in the concept that it could be used to Stabilize trauma. Trauma is something black men are all too familiar with. He states, " The purpose of this study is to explore the relationship between the impact of historical trauma and barriers on African-American males and the effects of emotional intelligence in reducing traumatic experiences." (Branford. D, 2020) The author informs the reader the research is based on previous research and studies that explore the historical review of African- American oppression, trauma in black males, and mental health in the African American community. He highlights his utilization of "historical trauma and emotional intelligence theories to explore barriers that African Americans have experienced over time and the role emotional intelligence can play in reducing trauma. It also explores the relevance of historical trauma and addresses opportunities for the implementation of emotional intelligence to improve mental health for African American males." (Brandford,2020). The irony that presents itself when reading and doing objective research is that while academic like Davis Brandford proposes the use of emotional intelligence in stabilizing trauma. Some debate whether emotional intelligence is even a valid construct. " Some personality psychologists argue that emotional intelligence can be more parsimoniously described by traits such as agreeableness and even charisma. A highly charismatic person, for example, is socially adept and can quickly read a room. While some studies have found a link between emotional intelligence and job performance, many others have shown no correlation whatsoever, and the lack of a scientifically valid scale makes it difficult to truly measure or predict how emotionally intuitive a person may be on the job or in other areas of life. Testing for EQ in the workplace, for example, is difficult because there is no validated psychometric test or scale for emotional intelligence as there is for the general intelligence factor—and many argue that emotional intelligence is, therefore, not an actual construct but a way of describing interpersonal skills." (Psychology Today, 2023)

What if IQ tests weren't what you thought they were? An even more ironic point is raised in Amos Wilson's "The Developmental Psychology of the Black Child" on page 141. Wilson educates on the development of the Mental Age (MA) concept by Binet attained by comparing the individual test performance with the average performance of a larger of individuals. Be essential procedure for constructing and standardizing intelligence tests of scales involves giving the proposed test to a large population and finding the ages at which a determined majority of these individuals have passed each item included in the test and arranging these items in order of difficulty. To calculate the MA of a particular individual, he is tested, and his results are compared to the average test performance of others his age. For example, if a child passed the test items passed by the average 7-year-old but not those passed by 8-year-olds, his MA would be seven; if his chronological age (CA) is also seven, then he may be said to be of average intelligence. If his CA is six, it may be said that he is above average intelligence. but if the CA happens to be 10 it may be said that he is below average intelligence." (Wilson. A, 2014) He goes on to highlight that German psychologist Wilhelm Stern developed the concept of intelligence quotient which can be Computed directly by using the formula 100MA/CA - IQ. It has been found that IQ is relatively constant for each individual. This should not be taken to mean that IQ is an absolute, Intrinsic, immutable quality of the individual and not subject to influence by several other factors and circumstances. IQ is more a measure of cognitive style related to social class, motivational state, and cultural background than it is to any innate biological capacity. Many variables affect the score of an individual and an intelligence test" (Wilson. A, 2014). He then proceeds to drop intellectual bombs stating, " up to this point, we have writtenOpen intelligence as if a universally accepted definition of this concept existed This is by no means the case. Guilford, a highly regarded psychometrist, spoke on the issue of definition: Aftertest has been invented to measure intelligence, quite several thinkers felt the urge to define it. Symposia were held on the problem, and numerous voices were heard. The outcomes were far from agreement. As spearmint1927 put it, intelligence became a mere focal sound, a word with so many meanings that it finally had none You further quoted JS Mill in a statement that describes the situation well, and that should serve as a warning Tennessee has always been strong to believe that whatever receives a name must be an entity of being having an independent existence of its own and if no real entity answering to the name could be found men did not for that reason supposed that none existed but imagine that it was something peculiarly abstruse and mysterious.

Despite EG Borings' somewhat facetious definition of intelligence as a measurable capacity that must, at the start, be defined as the capacity to do well in an intelligence test. There are serious questions as to what IQ tests to test and what is the structure of intelligence. Binet saw intelligence as a single unitary factor. Spearman, as a general factor plus several independent factors. Thurstone is a composite of simpler processes combined in complex ways to deal with the problems confronting the individual. Guildford posited a model for the structure of intelligence that consists of some 120 factors. From just these few examples, one can surmise that psychologists are far from agreement as to what intelligence is and what It is that intelligence tests are there for really measuring. The reader should keep this in mind when discussing intelligence or using the term to characterize some individual or group. It seems that the prime concern of psychologists in this area in with the problem of finding a universally scientifically accepted definition of intelligence has been misplaced towards The dubious area of trying to establish individual and ethnic group intellectual superiority or inferiority. Such exercises are patently ridiculous in light of the fact that the standard of comparison is undefined.." (Wilson. A, 2014)

So, if we are to entertain the debate around the validity of emotional intelligence, we would, by default, have to consider the lack of agreement in the area of the term intelligence. What Are the odds that the same validity issue they take with emotional intelligence is not taken with intelligence in general? At this point, I have to ask myself what purpose this position serves. Is the argument even worth entertaining? Wilson goes further, stating, " The fact that most IQ tests are standardized and involved the use of a large amount of statistical, mathematical formula and work that the IQ score is a number leave the impression on both lay and professional persons that these tests are objective other words psychosocial slides of

93

hand such as scientific methods of test construction and standardization and the representation of an individual's performance by some neutral number creates in the mind of the uninitiated and illusion of objectivity of intelligence test is largely mythologicalGinsburg listed four myths concerning the IQ test.

The first myth is that the IQ test measures intelligence which is a unitary mental ability. There is not one intelligence but many, and because of this is not clear what the IQ score reflects.

The second myth is that the differences in IQ scores reflect fundamental differences in intellect usual assumption is that differences in IQ reflect those abilities which are at the heart of the intellectual life. In addition, it is assumed that what the test fails to measure is not very important againThe proposition is in error.

The third myth is that the IQ test measures intellectual competence. The common view is that an individual's IQ reflects the best he can do in the intellectual spearThe IQ represents the upper limit of his mental capabilitiesWow this may be true for some people it is not for all the case of poor childrenespeciallyThe IQ test may not measure intellectual competence it may not give a true picture of what poor children are capable of.

The fourth myth is that the IQ test measures an innate ability that is relatively unaffected by the experience. In this view, the child's level of intelligence is set at birth, and later experience has relatively little effect on the IQ. This view is incorrectThe level of IQ is not determined at birth. It can fluctuate and change." (Wilson, 2014)

I find it very interesting that these myths and serious issues are negated in the attack on the validity of emotional intelligence. I believe more attention should be put on investigating emotional intelligence and its potential benefit on a personal level and a collective group level. Ultimately, my intent in choosing this topic to write on and publish is to plant this seed in the youth so they can pass it along and make a change. I don't just look up this information to

read at the moment and forget the next day. I learned this information and saw the work that needs to be done with my own kids as well as with myself. I pride myself in trying to be the best father, partner, and friend I can be, and consistent mindfulness of the state of my own emotional intelligence must be an obvious priority. I try to remind my kids when they are bickering and fussing to self-reflect and recognize not only their own emotions but the emotions of those around them. If Irahnae is upset because Zyon accidentally bumped, I get them both to acknowledge each other emotions as well as their own. Take them into consideration and apply them contextually. We work on impulse control and self-restraint more than anything in this house, and I have to be the model I want them to be. It is far from easy but well worth the reward of raising children with good character and integrity. Making emotional intelligence a priority and a major factor as a familial foundation is a highly probable way to attain that reward. Like the saying in the hood goes, "when you know better, you do better," and I live by that saying every day attempting to know something I didn't know before surpassing the person I was prior to not

knowing, therefore by default bettering myself. As I close this piece out, I feel like the perfect way to articulate my overall closing message and inspiration to put this publication together is to quote my big brother from another mother's publication that triggered it all " read study right research. Views research methodology Those are all what you need to become a revolutionary African scholar Listen to those who know more than you Challenge your mind to learn complex ideas Master those ideas so you can teach your people always inspire to push yourself to be the greatest scholar you can be always have a meticulous attention to detail mistakes are inevitable but correcting and recognizing mistakes is how you become a master teacher break ideas apart piece by piece learn and study each aspect and put those ideas back together compared to similar ideas and develop your own analysis from your own researchScarlett are made by how patient they are with new ideas that they struggle with the better you get under intellectual adversity the more you will grow and develop as a thinker We all have much more work to do Many more ideas to create new scientific discoveries waiting for us Research is a constant struggle but remember you aren't alone "As I learn we all learn sharing with others help deepen your own knowledge" (Piesie. K, 2021)

Chavis tp hsb warrior McCray.

CHAPTER SIX

Ethiopianism, Ethiopian Church & AME Church

Ethiopianism, Ethiopian Church & AME Church

By Kofi Piesie

What is Ethiopianism? Ethiopianism is a religious movement among sub-Saharan Africans that embodied the earliest stirrings toward religious and political freedom in the modern colonial period. The movement was initiated in the 1880s when South African mission workers began forming independent all-African churches, such as the Tembu tribal church (1884) and the Church of Africa (1889). An ex-Wesleyan minister, Mangena Mokone, was the first to use the term when he founded the Ethiopian Church (1892).

Nwadialor, Kanayo Louis, and Chukwuemeka Charles Nweke. "ETHIOPIANISM AND SOCIAL ECUMENISM: CHRISTIAN IDEOLOGIES FOR INDEPENDENT MOVEMENTS AND SUSTAINABLE NATIONAL INTEGRATION IN NIGERIA." ETHIOPIANISM AND SOCIAL ECUMENISM: CHRISTIAN IDEOLOGIES FOR INDEPENDENT MOVEMENTS AND SUSTAINABLE NATIONAL INTEGRATION IN NIGERIA, 3 July 2014, p. 17.

Ethiopia is located in the Horn of Africa, 3′ and 14.8″ latitude 33′ and 48′ longitude bordering

Somalia, the Sudan, Djibouti, Kenya, and Eritrea with a total border length of 5,311 km.

- It is the 10th largest country in Africa.

- The 2nd-most populous country in Africa after Nigeria

- 100 mts. Below sea level lowest pt

- 4620 mts. the highest pt. Ras Dashen

Its proximity to the Middle East and Europe, together with its easy access to the major ports of the region, enhances its international trade.

The most famous Ethiopian river is the Blue Nile or Abbay, which flows a distance of 1,450 kilometers from its source to join the White Nile at Khartoum. From the north and running down the center are the Abyssinian highlands. To the west of the chain, the land drops to the grasslands of the Sudan and to the east to the deserts of the Afar. South of Addis Ababa, the land is dominated by the Rift Valley Lakes.

"Overview about Ethiopia - Embassy of Ethiopia." Embassy of Ethiopia -, 7 Dec. 2020, https://ethiopianembassy.org/overview-about-ethiopia/#:~:text=Location%3A%20Ethiopia%20is%20strategically%20located,10th%20largest%20country%20in%20Africa.

Ethiopia is the oldest independent country in Africa and one of the oldest in the world. What are believed to be the oldest human remains.

I do not like the word Ethiopia because we do not have agency over the word Ethiopia or Ethiopian, and those people on that land mask did not refer to themselves as Ethiopian.

Origin of The Word Ethiopia

Ethiopian (n.)

late 14c., from Latin Æthiops "Ethiopian, negro," from Greek Aithiops, long supposed in popular etymology to be from aithein "to burn" + ōps "face" (compare aithops "fiery looking," later "sunburned").

"Ethiopian: Search Online Etymology Dictionary." Etymology, https://www.etymonline.com/search?q=ethiopian&ref=searchbar_searchhint.

According to Dr. Ayeke Bekerie, who is an Ethiopian-American Scholar and associate Professor and coordinator of the Ph.D. program in heritage studies as well as coordinator of the international affairs department at Mekelle University's Institute of Paleo-Environment and Heritage Conversation says on page 110

The Name Ethiopia invokes a unique identity incongruent with Africa. The name presumably was coined by outsiders, who lavished Ethiopians with praises and blamelessness, thereby, in the process, extrapolating an eternal source for the origin of Ethiopia's history and culture. According to an Ethiopian tradition, the term Ethiopia is derived from the word Ethiopis, a name of an Ethiopian king. The Ethiopian Book Askum identifies Ethiopis as the twelfth king of Ethiopia and the father of Aksumawi. Ethiopia's rich, varied, and original history forms a tangible center, not only in the self-naming of the Ethiopian but also in Affirming Africa's contribution to the world history. In the morning of the world, when the fingers of love swept aside the curtains of time, our dusky mother, Ethiopia, held the stage. It was she who wooed civilization and gave birth to nations. Egypt was her firstborn.

Bekerie, Ayele. "Ethiopica: Some Historical Reflections on the Origin of the Word Ethiopia." Journal of Ethiopian studies 1 (2016): n. page.

Despite Ethiopian traditions claiming a native origin for the country's name, many historians believe the word "Aethiopia" to be of Greek origin. Greek historian Herodotus used the word to refer to parts of Sub-Saharan Africa that were known to the Greeks at the time, specifically all inhabited land south of Egypt. It had also been used as a vague term for dark-skinned people since the time of Homer.

Reading George Hatke publication Aksum and Nubia: Warfare, Commerce, and Political

Fictions in Ancient Northeast Africa states on pages 52-53 that the reason why the territory of modern-day Ethiopia today claims this name may be due to the conquest of Meroe by the Axumite empire (located in modern-day Ethiopia and Eritrea) in the 4th century A.D., after which the Axumites began referring to themselves as "Ethiopians." This was likely due to the Biblical usage of the word "Ethiopian" and a desire for the newly Christianized Axumites to form a connection with Biblical tradition.

Hatke, George. Aksum and Nubia: Warfare, Commerce, and Political Fictions in Ancient Northeast Africa. NYU Press, 2013.

I am not trying to challenge Dr. Ayeke Bekerie, who is an Ethiopian-American Scholar and Associate Professor, about the term Ethiopia is derived from the word Ethiopis, a name of an Ethiopian king, but we do know that foreigners coined the name.

The picture on the right of the page is Mangena M. Mokon, born in the Transvaal in 1851. Mangena M. Mokone worked on a sugar plantation in Natal. He later moved to Durban, where he worked as a laborer during the day while attending night school. In 1874 he was baptized a Christian. The following year he began theological studies in Pietermaritzburg, Natal. He studied there for five years and was appointed to a preaching circuit of the Wesleyan Methodist Church in Natal. In 1882 he was posted to Pretoria, where he was to play a key role in the founding of the Kilnerton Training Institution, a secondary and post-secondary institution for the training of African male and female youth.

What cause rev. Mangena M. Mokone split with Methodist Church and organized a new church called the Ethiopian Church? It was Discrimination, Segregation, and Racism.

On November 1, 1892, Rev. Mangena M. Mokone, then an elder of the Wesleyan Methodist Church in Pretoria, severed his connection with that denomination. The causes for this action date back over a period of six years. Up to 1886, the white and African ministers in the Wesleyan Methodist Church met together in their district meeting. That year, the color line was drawn, and each side was required to meet apart from the other; yet the black ministers were compelled to have a white chairman and secretary.

Khoapa, Bennie A. "Mokone, Mangena Maake (b)." Dictionary of African Christian Biography, https://dacb.org/stories/southafrica/mokone-mangena2/.

Rev. Mangena M. Mokone resented this practice, and when no action was taken to correct the situation, he organized a new Ethiopian Church, with about 50 members, in Pretoria on Sunday, November 20, 1892. The name of the church was a biblical reference: "Ethiopia shall soon stretch out her hands unto God" (Psalm 68: 31). The first meeting was held in the Marabastad "Native Location" in an old, thatched house belonging to William Makando, who, although a Wesleyan minister, was in great sympathy with the Ethiopian Church Movement.

Khoapa, Bennie A. "Mokone, Mangena Maake (b)." Dictionary of African Christian Biography, https://dacb.org/stories/southafrica/mokone-mangena2/.

In January 1893, the Transvaal government recognized the Ethiopian Church. On

November 5, 1893, the Ethiopian Church was formally opened in the Marabastad location. The first preachers ordained in the Ethiopian Church were J.G. Xaba, ordained in 1894, and J.Z. Tantzi, ordained in 1895. They were both ordained by Rev. M.M. Mokone, assisted by Rev. J.M. Kanyane of an "Independent Church" known as the "African Church."

On May 31, 1895, Rev. Mangena M. Mokone, having heard about the African Methodist Episcopal Church (AME) in America, wrote to the bishop of that church seeking more information. With that information, he was impressed with what he heard about the AME Church in America.

AME African Methodist Episcopal Church

The AMEC grew out of the Free African Society (FAS), which Richard Allen, Absalom Jones, and others established in Philadelphia in 1787. When St. George's MEC officials pulled blacks off their knees while praying, FAS members discovered just how far American Methodists would go to enforce racial discrimination against African Americans. Hence, these St. George's members planned to transform their mutual aid society into an African congregation.

Ncurrie. "Richard Allen and the Origins of the AME Church." National Archives and Records Administration, National Archives and Records Administration, https://rediscovering-black-history.blogs.archives.gov/2021/06/09/richard-allen/.

Hmm, this sounds familiar right racial discrimination practices among Methodist Church, but we dig further into this in the chapter.

Who is Richard Allen?

Richard Allen was born February 14, 1760, enslaved to Benjamin Chew, a Quaker lawyer in Philadelphia. As a child, he was sold to Stokley Sturgis, a plantation owner in Dover, DE, where Allen taught himself to read and write. Allen's owner was involved in the Methodist Church and permitted his slaves to attend their services. Allen was also drawn to the Church and began evangelizing as a teenager while still enslaved.

In 1780 he bought his freedom for $2000 (approximately $36,364 in modern U.S. dollars) by working on his own time over a period of five years. He continued his involvement with the Church, and in 1786 he became a preacher at St. George's Methodist Church but was only permitted to conduct segregated sermons. The frustration of segregation and further racial tensions led Allen and more than forty others to leave St. George's

Although most wanted to affiliate with the Protestant Episcopal Church, Allen led a small group that resolved to remain Methodists. In 1794 Bethel AME was dedicated with Allen as pastor. To establish Bethel's independence from interfering white Methodists, Allen, a former Delaware slave, successfully sued in the Pennsylvania courts in 1807 and 1815 for the right of his congregation to exist as an independent institution. Because black Methodists in other Middle Atlantic communities encountered racism and desired religious autonomy, Allen called them to meet in Philadelphia to form a new Wesleyan denomination, the AME.

The geographical spread of the AMEC before the Civil War was mainly restricted to the Northeast and Midwest. Major congregations were established in Philadelphia, New York, Boston, Pittsburgh, Baltimore, Washington, DC, Cincinnati, Chicago, Detroit, and other large Blacksmith's Shop cities. Numerous northern communities also gained a substantial AME presence. Remarkably, the slave states of Maryland, Kentucky, Missouri, Louisiana, and, for a few years, South Carolina became additional locations for AME congregations. The denomination reached the Pacific Coast in the early 1850s with churches in Stockton, Sacramento, San Francisco, and other places in California. Moreover, Bishop Morris Brown established the Canada Annual Conference.

As you can see, the AME churches begin popping up in the East, Southeast, West, Midwest, and South states. AME community work grew the popularity of the church. Its new ideal African for Africans was a great start for the church, which reached all the way to South Africa.

Unification of Ethiopian Church and (AME) African Methodist Episcopal Church

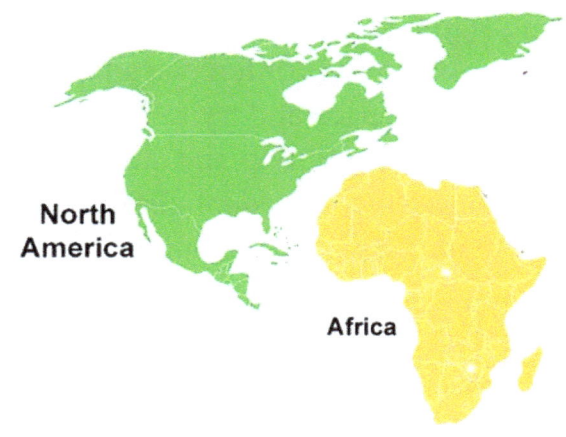

Rev. Mangena M. Mokone persuaded his followers to affiliate there. Consequently, at the third session of the Ethiopian Church Conference held in Pretoria on March 17, 1896, it was resolved to unite with the African Methodist Episcopal Church. At that conference, the Rev. James Dwane, a Wesleyan minister who was destined to become an outstanding figure in the Ethiopian Church in South Africa, joined the conference and was elected to go to America and try to consummate the union. In 1898 the Ethiopian Church became the fourteenth episcopal district of the African Methodist Episcopal Church.

Khoapa, Bennie A. "Mokone, Mangena Maake (b)." Dictionary of African Christian Biography, https://dacb.org/stories/southafrica/mokone-mangena2/

Who is James M. Dwane?

James M. Dwane (1848-1915), born in the Cape Colony, was one of the African church leaders who, in 1892, founded the independent Ethiopian Church of South Africa, a powerful movement that linked Christianity with political action.

Originally a leader in the Wesleyan Mission Church, Dwane opposed the racial segregation he found practiced within the white-controlled church. He had also quarreled with the Wesleyan authorities over the expenditure of funds he had raised for the church during a visit to England. He resigned from the Wesleyan Church in 1892 and proceeded to establish the Ethiopian Church with other dissatisfied church members.

An eloquent speaker, the Rev. Dwane became interested in the African Methodist Episcopal (AME) Church founded by African Americans in the United States in 1816. It was decided at a conference of all independent church leaders held in Pretoria in 1896 that the Rev. Dwane and two other African churchmen, one of whom was Mangena Mokane, should go to the United States to obtain affiliation with the AME Church. Only Dwane, however, was able to raise funds for the journey. He sailed for Philadelphia in 1896. Dwane presented his case well in America, and the Ethiopian Church of South Africa was formally incorporated in the African Methodist Episcopal Church. Before his return to Africa, Dwane was appointed general superintendent of the South African member church.

What did the Africans in South Africa and The Africans in America have in common with the church? Discrimination, Segregation, and Racism.

Parallel developments occurred elsewhere and for similar reasons. In Nigeria, the so-called African churches—the Native Baptist Church (1888), the former Anglican United Native African Church (1891) and its later divisions, and the United African Methodist Church (1917)—were important. Other Ethiopian-related movements were represented by the Native Baptist Church (1887) of Cameroon; by the Native Baptist Church (1898) in Ghana; in Rhodesia by a branch (1906) of the American Negro denomination, the African Methodist Episcopal Church, and by Nemapare's African Methodist Church (1947); and by the Kenyan Church of Christ in Africa (1957), formerly Anglican.

<small>Adejumobi, Saheed. "Ethiopianism •." •, 15 July 2020, https://www.blackpast.org/global-african-history/ethiopianism/.</small>

Saheed Adejumobi goes on further in his article Ethiopianism that early Ethiopianism, which included tribalist, nationalist, and Pan-African dimensions, was encouraged by association with independent American black churches and radical leaders with "back to Africa" ideas and an Ethiopianist ideology. This ideology was explicit in the thought of such pioneers of African cultural, religious, and political independence as Edward Wilmot Blyden and Joseph Ephraim Casely-Hayford of Ghana (e.g., his Ethiopia Unbound, 1969).

Adejumobi, Saheed. "Ethiopianism •." •, 15 July 2020, https://www.blackpast.org/global-african-history/ethiopianism/.

Conclusion: Significance

The movement, just as initially perceived by the proprietors, did serve its purpose to a great extent. It saw that Africans were liberated from harsh treatment by colonial leaders and ensured that retrogressive issues such as racial segregation were going into extinction. The movement played a key role in helping the Zulu rebellion become a great success in 1906 under the leadership of John Chilembwe. The movement also saw that Africans could now take up leadership positions, especially in the churches putting them in a position to make influential decisions. The movement ensured that the slogan "Africa for Africans" came to pass with full inclusivity across the board; religion, political as well as social setup in the colonial era. The movement facilitated the promotion of nineteenth-century, and later, Pan-African nationalism, which transcended Western-imported denominationalism and united black people.

CHAPTER SEVEN

Examining available data on the Pros and cons of eBooks versus physical paperback books. Should we still write in an era of smartphones and iPads?

Examining available data on the Pros and cons of eBooks versus physical paperback books. Should we still write in an era of smartphones and iPads?

Chavis Tp hsbAhaw McCray

This chapter is inspired by a presentation I did on the Kofi Piesie YouTube Channel in April of 2022. At the time we were focused as a research team to make sure that we weren't getting too caught up in the same repetitive topics as other channels and I sought to personally go in a completely different direction.

Much of the talk on other channels lacked in the area of reading comprehension and it just so happened a scientific Nature article I read called "Reading on a smartphone affects sigh generation, brain activity and comprehension" had recently been published.

The article was right on time and very intriguing because we live in an era where everyone is reading things on their phone vs buying and reading physical books if they are even reading at all. What I read eventually sparked an interest to investigate further and lead to the presentation. I received a lot of positive feedback from it so I felt it was worthy enough to go ahead and put in a book seeing as how

this book targets the youth and knowledge seekers in general. I feel like whoever I may miss with the presentation I may reach through this publication. More than anything I feel this type of information is perfect as it deals with a modern issue that most intellectuals may overlook. 20 years ago we may not have even thought of this because we didn't even have smartphones vs today where 9-year-old kids have iPhones and schools recently had to teach kids to use online platforms because of the pandemic.

My kids did well with technology to learn because we thankfully had bought them phones and I am always on them about using them to learn. Others aren't always able to afford all these devices and I felt like my kids being familiar would give them a leg up in their intellectual journey. A sociology course I took touched on how societies today that were technologically advanced usually tend to be more advanced than societies that lacked in their area. This article raised my eyebrows when I dug into it because its conclusion made me start asking a lot of questions because what

I found partially challenged the idea contextually.

My objective was to assess multiple scientific journal studies and identify the pros and cons of reading Ebooks and physical paper books. I also aimed to evaluate the necessity of writing, and its significance or irrelevance in a world where technology is king and smartphones and computers rule. The research questions I chose went as follows: What are the pros and cons of reading Ebooks? What are the pros and cons of reading physical paperback books? Should we even write in an era where everything is being typed up on iPads and smartphones? Though it may sound remedial the first thing I like to do is define terms and I started with a PRINTED BOOK. "A printed book is a literary publication comprising of pages bound together along a single side and, protected by a cover." (Surbi, 2020) I then proceeded to define eBooks which are defined as "a book that is transformed into electronic form, for reading on a dedicated e-reader or computer and handheld devices, is called an eBook." (Surbi, 2020) After defining the terms I wanted to identify the differences between the 2 which is the fact one is read off a screen while the other is read off paper negating the reality of physical storage vs digital storage. Other

differences include physical damage and loss of physical books vs not being able to damage a

digital file and easy searches for recovering E-books. E-books are also usually less expensive than physical books. EBooks can't be resold vs physical copies are resold regularly.

Outline of that I 1st wanted to identify the pros or positives associated with reading ebooks. I wanted to see what scientifically had been assessed so scientific journals and articles were the source material that I had to review. One of the first articles I reviewed and presented was titled "Impact of tablet computers and eBooks on learning practices of law students". I saw it as relevant to this discussion because it involved law students learning via the use of these devices to read eBooks on them. In regards to Ebooks, the following was worth citing, " as a mobile, connected platform, eBooks have the unrealized potential to support more pedagogical approaches than traditional books, including experiential, problem-based, dynamic, and social learning. (Kalz et al, 2012)

If pedagogical is a new term for you the root of this term is pedagogy (ped·a·go·gy) which is defined as "the art, science, or profession of teaching."(https://www.merriam-webster.com/dictionary/pedagogy) So based on the article research, eBooks support more ways of teaching than traditional books. The authors add, " To realize this potential, the Personalized eBooks for Learning (PEBL) the project has developed a specification that enables new capabilities in eBooks while maintaining the advantages of the "book" format. The PEBL project funded by the US Advanced Distributed Learning (ADL).

The initiative has extended the standard EPUB3 format to enable eBooks to communicate with other systems in live, virtual, simulated, and constructive environments; to embed and exchange data with simulations, games, and intelligent tutoring systems; and to serve as competency-based training environments."(Kalz et al., 2012) So basically, Ebooks' advantages over traditional books were they could communicate with other systems to teach in ways a traditional teacher with a traditional book could only imagine due to the technology of the device in this article.

In another article titled "Interactive features of E-texts' effects on learning: a systematic review and meta-analysis" the author's abstract highlights "Based on the meta-analysis, interactive features benefited reading performance (g = .66, p < .001). Individual studies with positive effects involved multiple interactive features; however, potential contributions of three types of features (questions with feedback, digital glossaries, and collaborative tools) are discussed."(Clinton-Lisell,, 2020) So this meta-analysis analysis essentially demonstrates how features on these devices were a benefit to reading performance because the features offer not only questions for feedback but other tools to help navigate and increase knowledge.

I thought this 2022 paper titled "What is the effect of touchscreen technology on young children's learning?: A systematic review" offers some more evidence of positive benefits through the majority of the systematic review discovered that they "generally advocated positive effects of touchscreen devices on young children's learning with 34 studies reporting positive effects." (Kalati, A.T and Kim, M.S, 2022) The authors also uncovered "17 studies obtaining mixed findings, and 2 articles reporting negative effects" (Kalati, A.T and Kim, M.S, 2022) Overall the authors of this study concluded "findings has implications for different stakeholders by giving them insights into the impact of touchscreen devices on young children's learning under different conditions." (Kalati, A.T and Kim, M.S, 2022)

Lastly, before I move on to the cons I wish to present another study in which the authors state "Research suggests that technology use among adult literacy students has positive effects on learning outcomes, with the majority of these studies focused on traditional classroom environments where technology is used as a supplement to existing instruction" (Milheim. K, 2007) Though this was positive if we critically think we see that maybe technology use isn't all that great for our fundamentals because we see they are better as supplements for traditional and foundational instruction that we get from books like the one your reading.

To start looking at the cons I think now is a good time to bring up the Nature article and its finding where it states that " One disadvantage is that reading comprehension is reduced when reading from an electronic device; the cause of this deficit in performance is unclear.

In this study, we investigated the cause for comprehension decline when reading on a smartphone by simultaneously measuring respiration and brain activity during reading in 34 healthy individuals."(Honma et al, 2022) The authors elaborate further when they state that they "found that, compared to reading on a paper medium, reading on a smartphone elicits fewer sighs, promotes brain overactivity in the prefrontal cortex, and results in reduced comprehension. Furthermore, reading on a smartphone affected sigh frequency but not normal breathing, suggesting that normal breathing and sigh generation are mediated by pathways differentially influenced by the visual environment. A path analysis suggests that the interactive relationship between sigh inhibition and overactivity in the prefrontal cortex causes comprehension decline."(Honma et al, 2022)

This to me was a significant strike against eBooks. This article hit home because me and my good brother Shawn of KPRT and Mossi Warrior Clan had been going back and forth about the importance of reading and writing and he essentially took the position that was lazy for not writing out my surveys for our volume 4 publication. I took the position that I was just being innovative because I took notes on my phone and typed up/talk to text my submissions instead of writing them down in notepads. I essentially focused on all the pros of technology that I ironically used here except I hadn't yet done this research. I conceded to being lazy but I held on to my innovative position as I argued "it was 2022 we got smartphones now, what do I look like not taking advantage?". Well, I look like someone who may be stunting my comprehension according to the evidence presented by this rather humbling study.

While assessing the article I found myself skimming through the author's sources and found an interesting article titled "Effects of VDT and paper presentation on consumption and production of information: Psychological and physiological factors". VDT was an acronym for Visual Display Terminals. This article's results "show that performance in the VDT presentation condition was inferior to that of the Paper presentation condition for both consumption and production of information. Concomitantly, participants in the VDT presentation condition of the consumption of information study reported higher levels of experienced stress and tiredness whereas the participants in the VDT presentation condition of production of information study reported only slightly higher levels of stress." (Waslund, E and Archer.T, 2005) This also hit home personally as I am an owner of multiple pairs of glasses and my vision is extremely poor already. This study was saying essentially not only were there issues with the consumption of the information on these devices but issues with the production of it.

It was saying there were issues about fatigue and computers overworking your brain because so much was going on that we never stop to think about. To lend more validation to this a study titled "Usability evaluation of E-books" highlights how "reading an E-book causes significantly higher eye fatigue than reading a C-book. Reading a C-book generated a higher level of reading performance than reading an E-book. In addition, females demonstrated better reading performance than males in reading either book."(Kang, Y.Y and Lin, R, 2009) More emphasis on lighting, font size, and other aspects of the display are discussed in a paper titled "Lighting, font style, and polarity on visual performance and visual fatigue with electronic paper displays" and the point of the device's lighting needing greater illumination was a major highlight that took another jab at eBooks.

When we look at paper medium pros we find reading on a paper medium doesn't elicit fewer sighs or promote overactive Brain activity which the nature article asserts a link between these conditions and a decrease in reading comprehension.

The paper medium doesn't utilize harmful lights that aren't healthy for your eyes.

Paper medium proved superior in the area of consumption and production of information. No special light is necessary for paper mediums. The paper medium can be damaged or misplaced. Paper mediums aren't as innovative as ebooks in regards to being integrated to communicate with other systems in live, virtual, simulated, and constructive environments; to embed and exchange data with simulations, games, and intelligent tutoring systems; and to serve as competency-based training environments. Eventual storage and waste issues will be experienced with paper medium.

Searches are less convenient via a paper medium. Physical books are more expensive than ebooks. Some physical copies of books go out of print. Physical books are easier to steal.

An interesting find in comparison of comprehension tests administered by pen and pencil vs computer found no real advantage either way. (Margolin et al, 2013) Another study indicated "Although reading comprehension and writing are different tasks, research indicates that they overlap in terms of their purposes, processes, and sources of knowledge." (Allen. L et al, 2014) This is contextually emphasized in a 2011 meta-analysis emphasizing the significance of reading as well as writing stating that "Graham and Herbert present evidence that writing about material read improves students' comprehension of it; that teaching students how to write improves their reading comprehension, reading fluency, and word reading; and that increasing how much students write enhances their reading comprehension. These findings provide empirical support for long-standing beliefs about the power of writing to facilitate reading" (Graham, S & Herbert, M, 2011)

This earlier study supports this overlap but doesn't it necessarily support reading ebooks? Could one write about what they read in an ebook? Sure, but what are we writing on? Most of us write on PAPER where we would then READ IT to consume it. I take notes on my phone which is a device with a visual display terminal. To be completely honest my ability to verbatim recall this information is poor I just document the info in my notes like a bookmark where I know the information is located in a particular spot for me to recall. It would benefit me to write these notes by hand to commit them to memory better. Another article echoes this sentiment

"Reading comprehension is a complex process. To understand a text, the reader needs to recognize its words and access their meaning, the reader needs to activate related knowledge must be activated, and the reader needs to generate inferences as information is integrated during the time of reading. Thus, students' writing is affected by their reading, and how they understand what they have read."(Fahad. A, 2015)

In conclusion, the pros and cons vary for both Ebooks and Paper mediums. For Ebooks, the pros range from being better in the storage department to convenient searches while its cons range from being harmful to the eyes to decreased reading comprehension. For paper medium increased reading comprehension was a constant pro while storage, potential damage, and less convenient searches were obvious cons. Writing is also linked to improved reading comprehension, fluency, and word fluency based on available data so we should incorporate a lot of writing if we won't fully grasp what we read. New technology supplements traditional learning but our foundational skill set is sharpened by what some may feel is outdated means of obtaining knowledge through reading physical books based on that and the available evidence I think we should still read physical books vs throwing them away and collecting ebooks. There is a give and take with anything but take away your foundation and it becomes almost impossible to build.

CHAPTER EIGHT
An interesting Message from AI

An interesting Message from AI

Chavis Tp hsbAhaw McCray

This chapter is pretty much inspired by and majority composed by Artificial intelligence. I had the idea to write on the topic of desensitization to violence concerning the African American community and never really finished. The outline I came up with looked like this:

1. What is desensitization?

2. Should black people be concerned about the state of desensitization to violence in their community? Why?

3. Does media influence increased desensitization?

4. Is there a link between internalization and desensitization in kids?

5. How does living in a dangerous environment affect kids?

6. What type of impact could this have on black kids specifically?

7. Is disrespect of the dead and drill music associated with black people's desensitization to violence?

8. How can black people solve the issue of desensitization to violence or at that minimum cope?

Now in all honesty I thought I answered every one of these questions utilizing sources I did a literature review on to create these questions. For example to answer the first question defining the concept of discussion an article titled Violence in Films and Television by Joanne Cantor from the Encyclopedia of International Media and Communications in 2003 states: "Desensitization is a psychological process by which a response is repeatedly elicited in situations where the action tendency that arises out of the emotion proves to be irrelevant. Desensitization is sometimes used to treat phobias by gradually and repeatedly presenting the frightening stimulus under non-threatening conditions. Over time, when desensitization works, the phobic response becomes less and less intense" (Cantour. J, 2003)

I planned to form my argument to answer question 2 but got distracted with life and multi-tasking but was interestingly intrigued weeks down the line when I was scrolling my Facebook timeline and saw my good brother Rakeem Shabazz's random post on how he used something called chatgpt to take a safety quiz in which it got him 9 out of 10 questions answered correctly. That stuck in my head as I would see another good brother who identifies as Divin Rulah that is heavy into Artificial intelligence make posts speaking on Chatgpt and its issues. I then watched a YouTube video from a prominent news station that gave the best and most interesting demonstration of what chatgpt was and how it was being used in schools by kids to trick teachers into thinking they wrote a paper when in reality they used AI on an app to write the paper. A high school teacher was tested to see if he could identify the cheaters and to my surprise, he could not. This led me to look into this chatgpt for myself.

If you look on the website you are informed that Chatgpt is where programmers "trained a model called ChatGPT which interacts in a conversational way. The dialogue format makes it possible for ChatGPT to answer follow-up questions, admit its mistakes, challenge incorrect premises, and reject inappropriate requests. ChatGPT is a sibling model to

InstructGPT, which is trained to follow an instruction in a prompt and provide a detailed response." (OpenAI, 2022)

I had never heard of this before and quickly downloaded the app called chat sonic which helps write on topics. The app prompts you to ask what it is you want to write about. I asked could it help me write about desensitization to violence and African American youth and to my amazement in less than 20 seconds this is the response I was given that pretty much ironically answered the remainder of the questions I sought to answer from the outline :

" Absolutely! Desensitization to violence among African American youth has become an increasingly concerning issue in recent years, as it has been found to be a catalyst for violent behavior and a serious threat to the well-being of young people of color.

Studies have shown that African American youth are exposed to more violence in their everyday lives, both through direct experience and by consuming media such as television, movies, and video games. This increased exposure to violence desensitizes them to its effects, leading to a diminished sensitivity to the moral and emotional consequences of their actions.

The lack of awareness of the potential harm that can come from engaging in violent behavior further contributes to the problem of desensitization. This is especially true for African American youth, who are often exposed to violence at a much younger age and are thus more likely to have a decreased sense of empathy and understanding of the impact of their decisions.

Various interventions have been proposed by researchers and practitioners in order to address the problem of desensitization to violence among African American youth. These include the use of therapy, mentoring programs, and other forms of psychoeducation to help young people develop a better understanding of the consequences of their decisions. Additionally, increasing access to community programs, recreational activities, and other outlets that promote positive behavior can also serve as a way to counter the effects of desensitization.

Ultimately, desensitization to violence among African American youth must be addressed through a combination of education and access to resources. By providing young people with the knowledge and tools needed to make better decisions, we can help reduce the negative impact of desensitization and ensure that everyone is given the same opportunity to thrive" (ChatSonic, 2022)

As you can see this response was extremely detailed and spot on. Pretty much everything I wanted to cover was addressed by artificial intelligence in seconds. My purpose in making this book worthy is to not just show AI can help write papers and essays but to trigger thought as to the direction of our current state of technology. Who knows how many times or how much of some of the stuff we read today is AI generated versus independently composed? I encourage the reader to make themselves familiar with the current state of technology not so they can cheat on homework and test but so they don't get left behind. Find innovative ways to utilize it for your community and personal life's benefit. Hopefully, this information inspires curiosity in technology and motivates the reader to become acquainted with AI.

Jama Mboka (Peace Family)

CHAPTER NINE

Introduction into Authorship

Introduction into Authorship

Chavis McCray Jr/ Khayrii McCray/ Landon Garcia-McCray

In this section of the book, I challenged my nephew and sons to write about things they were essentially passionate about wanting others to know so I could add them to this publication and make them grade school authors part of their first publication. Being that I'm in the position to make their wishes of being in a book a reality here is what they wrote:

Muhammad Ali

" Born and raised in Louisville, Kentucky, he began training as an amateur boxer at age 12. At 18, he won a gold medal in the light heavyweight division at the 1960 Summer Olympics and turned professional later that year. "

(https://en.m.wikipedia.org/wiki/Muhammad_Ali)

He was very intelligent in his moves and he was a person who never gave up. He was inspiring to me and a lot of others. I am a person who doesn't like to give up too. I feel like we are just alike because he never would give up even if it is a life or death situation just like me.

He became a Muslim after 1961. He won the world heavyweight championship, defeating Sonny Liston in a major upset on February 25, 1964, at age 22. During that year, he denounced his birth name as a "slave name" and formally changed his name to Muhammad Ali.

People have been telling me I should give up on what I love but when I think about Muhammad I'm inspired and I don't care what they say and I keep on doing what I love and that is Football.

This type of attitude has been so important to me and my future because I'm going to keep on getting hurt and I will not always be the best and Muhammad Ali tells me s "I hated every minute of training, but I said, 'Don't quit. Suffer now and live the rest of your life as a champion.'" Muhammad Ali

Chavis McCray Jr aka Cj Macc

1st Football Experience

What made me interested in playing football is watching people make awesome catches and jukes. I also thought it will make me better when I get to the next level. Everything is that it will teach me about sportsmanship and teamwork. It also teaches me how to accept a win or a loss. Lastly, it will teach me how to keep my composure.

The first practice was difficult. We ran laps I got very tired quickly. Barely knew the basics of football. I thought football was all about fancy catches and jukes. But then my coach pulled me to the side and told me every play does not have to be a highlight play.

When it was my first game I was very excited. I was warming up hoping I will get in the game. Our captains went to see if the kickoff return will get on the field or kick off or get on the field. I was ready to use the tips the coach gave me like catching the ball getting positive yards and tackling low. My first game was good but we lost and I was very upset.

After the game coach taught me how to accept my losses like I would do my wins. During the next couple of practices, I got very good. I also got the starting wide receiver spot. I also learned this play called a jet sweep. In the next

We couldn't make it to the super bowl playoffs but there's this thing called the copper bowl. Game of the copper bowl playoffs we won but sadly we lost to the cougars which set us home. But most of all I learned about sportsmanship and how to be a man that's all that matters. Not winning trophies and rings. Although those things are nice they won't help you as a human being. Another thing about football is it helps you build a relationship with your team as if you were brothers. Like my coach says there is nothing like a brother's love.

Khayrii McCray

Desmond Demound Bryant was born on November 4th, 1988. He was an American football player who was a wide receiver for my favorite team the Dallas Cowboys in the National Football League. He played college football for Oklahoma State where he earned All-American. In 2008 he was selected by the Dallas Cowboys and the first round of the 2010 NFL draft. He earned three Pro Bowl appearances and also was named all-pro in 2014. Something he said that sticks with me that I heard him say is "to be a man and respect people."

Landon Garcia-McCray

Sources

Alexander, Michelle (The New Jim Crow: Mass Incarceration in the Age of Colorblindness, TheNew Press, 2010)

Jr. Bennett, Lerone (Before The Mayflower 6 th Revised Edition, Penguin Books, 1963)

Davis, J. Angela (Policing the Blackman, Penguin Random House, 2017)

Garner, Bryan (Black's Law Dictionary 5 th pocket edition, 2016)

Hall, Kermit (The Oxford Companion To The Supreme Court Of The United States, Oxford University Press,1992)

Jordan, Terry (The U.S. Constitution and Fascinating Facts About It, Oak Hill Publishing, 2016)

ONLINE REFERENCES

https://www.britannica.com/nathanbedfordforrest

https://britannica.com/warondrugs

https://education.nationalgeographic.org

https://www.pbs.org

https://en.m.wikipedia.org/newyorkslaverevolt

https://en.m.wikipedia.org/warondrugs

Ashley G. Chrysler, Note, Lyrical Lies: Examining the Use of Violent Rap Lyrics as Character Evidence

Under FRE 404(b) and 403 11 (2015) (published online by Michigan State University College of Law),

[https://perma.cc/VT88-7KXP].

https://www.ajc.com/news/crime/young-thug-other-defendants-facing-new-charges-in-gang-

indictment/G2TL6B27NJFCJMOZIJCLTZPCS4/

https://www.thetexastrialattorney.com/blog/miranda-warning-

texas/#:~:text=If%20you%20were%20not%20read,cannot%20be%20used%20against%20you.

https://www.law.cornell.edu/rules/fre/rule_404

https://jgrj.law.uiowa.edu/online-edition/volume-24-issue-1/bars-behind-bars-rap-lyrics-character-evidence-and-state-v-skinner/

https://www.nme.com/news/music/new-york-senate-passes-bill-limiting-use-of-song-lyrics-as-evidence-in-court-3228154

https://www.nbcnews.com/news/nbcblk/georgia-da-fani-willis-says-rap-lyrics-will-continue-used-criminal-cas-rcna45680

https://constitution.congress.gov/constitution/amendment-1/

https://www.ajol.info/index.php/gjrt/article/view/227818

Benjamin Bannekar letter to Thomas Jefferson

https://founders.archives.gov/documents/Jefferson/01-22-02-0049

Title: The Souls of Black Folk Author: W. E. B. Du Bois Release Date: January, 1996 [eBook #408][Most recently updated: August 11, 2021]

African Pygmy Vertical Learning processes: https://www. ecoanthro.net/vertical-learning-systems/

The Role of Traditional Knowledge in African Pygmy Culture: https://www. researchgate.net/publication/304158423_The_Role_of_Traditional_Knowledge_in_African_Pygmy_Culture

The Value of Vertical Learning in African Pygmy Communities: https://www.intechopen.com/books/rise-of-indigenous-knowledge/the-value-of-vertical-learning-in-african-pygmy-communities

Vertical Learning Among the African Pygmies: https://www.semanticscholar.org/paper/Vertical-Learning-Among-the-African-Pygmies-Burke/21e7e7512d6e5c6d17235183f5b5e5a3d9c96f03

African Information Systems Knowledge sharing networks The family Unit being the most vital component information sharing,

Conservation in Indigenous communities

https://www.scientificamerican.com/article/what-conservation-efforts-can-learn-from-indigenous-communities/

The Concept of Health and Wholeness in Traditional African Religion and Social Medicine

https://www.researchgate.net/publication/327675390_The_Concept_of_Health_and_Wholeness_in_Traditional_African_Religion_and_Social_Medicine

Traditional African medicine

https://www.ncbi.nlm.nih.gov/pmc/articles/PMC3866779/

Indigenous Knowledge Systems for sharing information regarding medicinal plants

https://www.mdpi.com/1424-2818/14/3/192/htm

Focusing on traditional African Medicine in primary health care. https://www.ncbi.nlm.nih.gov/pmc/articles/PMC1447208/

African Traditional Medicine and Its Role in Conservation. https://www.ncbi.nlm.nih.gov/pmc/articles/PMC1360614/

Nigerian Traditional Herbal Medicine-The Treatment of Malaria. https://www.ncbi.nlm.nih.gov/pmc/articles/PMC5182145/

African Traditional Medicine and Healthcare Inequality in Sub-Saharan Africa: A systematic scoping review. https://journals.plos.org/plosone/article?id=10.1371/journal.pone.0236327

African Traditional Medicine Healing Practices: Community Knowledge and Attitudes in

Uganda. https://www.ncbi.nlm.nih.gov/pmc/articles/PMC5483171/

Use and Benefit Sharing for the Valorization of Traditional African Medicine. https://www.ncbi.nlm.nih.gov/pmc/articles/PMC6510292/

Pfizer, Merck Join Wave to Profit from Traditional African Medicines. https://www.harmalive.com/news/pfizer-merck-join-wave-to-profit-from-traditional-african-medicines

Negotiating the Ethics of Western Bioprospecting of African Medicinal Plant Diversity. https://www.ncbi.nlm.nih.gov/pmc/articles/PMC6440195/

Unbalanced Benefits between the African Traditional Medicine and the Western Pharmaceutical Industries. https://www.ncbi.nlm.nih.gov/pmc/articles/PMC6533616/

African Traditional Medicine and Pharmaceutical Patents. https://www.lexology.com/library/detail.aspx?g=887d2e27-89c4-4bcb-8456-ad0ebaa3bf3b

The codification of Indigenous African Knowledge systems

https://journals.sagepub.com/doi/abs/10.1177/0266666919853007

https://dl.acm.org/doi/abs/10.1145/3239264.3239266

https://www.ncbi.nlm.nih.gov/pmc/articles/PMC4803932/

https://files.eric.ed.gov/fulltext/EJ1017665.pdf

Capitalize on African biodiversity

https://www.nature.com/articles/548007a

African Secret Societies

"Secret Societies of West Africa" by William Bascom

"Secrets and Knowledge in Medicine and Science, 1500-1800" edited by Elaine Leong and Alisha Rankin

"Secret Societies in Nigeria: Their Functions and Influence on the Nigerian Society" by Olumide Olumilua

Cell Press. "Tracing the path of pygmies' shared knowledge of medicinal plants." ScienceDaily. ScienceDaily, 8 September 2016.

<www.sciencedaily.com/releases/2016/09/160908130558.htm>.

Africa's Traditional Knowledge System In Global Trade
https://papers.ssrn.com/sol3/papers.cfm?abstract_id=1613383

https://www.ncbi.nlm.nih.gov/pmc/articles/PMC5442578/#:~:text=Indigenous%20knowledge%20or%20African%20knowledge,deeply%20embedded%20in%20cultural%20values.

https://www.inst.at/trans/17Nr/9-3/9-3_sifuna.htm

Emotional Intelligence

https://www.psychologytoday.com/us/basics/emotional-intelligence

EMOTIONAL INTELLIGENCE IN AFRICAN AMERICANS: HOW CULTURE

AFFECTS THE APPLICABILITY OF THE CONSTRUCT

Kristen Y. Sanders, M.A. 2021

https://www.proquest.com/openview/02ce3f056f2404a1ee931d4c89093b42/1?pq-origsite=gscholar&cbl=18750&diss=y

Brandford, Davis, "The Mental Health of Black Men: Stabilizing Trauma with Emotional Intelligence" (2020). School of Professional Studies. 46.

https://commons.clarku.edu/sps_masters_papers/46

Piesie, Kofi, "As I learned, we all learned sharing with others helps deepen your own knowledge" (2021)

Wilson, Amos "The psychological development of the black child" (2014)

Effects of high pixel density on reading comprehension, proofreading performance, mood state, and

physical discomfort

https://www.sciencedirect.com/science/article/abs/pii/S0141938216301494

E-readers, Computer Screens, or Paper: Does Reading Comprehension Change Across Media Platforms?

https://onlinelibrary.wiley.com/doi/abs/10.1002/acp.2930

The Cognitive Equivalence of Reading Comprehension Test Items Via Computerized and Paper-and-

Pencil Administration

https://www.tandfonline.com/doi/abs/10.1207/S15324818AME1602_2

Reading comprehension components and their relation to writing

https://www.cairn.info/revue-l-annee-psychologique1-2014-4-page-663.htm

Writing for Learning to Improve Students' Comprehension at the College Level

https://eric.ed.gov/?id=EJ1075302

Reading linear texts on paper versus computer screen: Effects on reading comprehension

https://www.sciencedirect.com/science/article/abs/pii/S0883035512001127

Writing to Read: A Meta-Analysis of the Impact of Writing and Writing Instruction on Reading

Steve Graham; Michael Hebert

Harvard Educational Review (2011) 81 (4): 710–744.

https://doi.org/10.17763/haer.81.4.t2k0m13756113566

https://meridian.allenpress.com/her/article-abstract/81/4/710/32006/Writing-to-Read-A-Meta-Analysis-of-the-Impact-of?redirectedFrom=fulltext

The Relationship Between Visual Contrast Sensitivity and Neuropsychological Performance in a Healthy Elderly Sample

https://www.tandfonline.com/doi/full/10.1080/13803390590954173

Lighting, font style, and polarity on visual performance and visual fatigue with electronic paper displays

https://www.sciencedirect.com/science/article/abs/pii/S0141938208000838

Usability evaluation of E-books

https://www.sciencedirect.com/science/article/abs/pii/S0141938208000826?via%3Dihub

Effects of VDT and paper presentation on consumption and production of information: Psychological

and physiological factors

https://www.sciencedirect.com/science/article/abs/pii/S0747563204000202?via%3Dihub

Influence of Technology on Informal Learning

https://www.learntechlib.org/p/71829/

A systematic review of research on e-book-based language learning

http://www.kmel-journal.org/ojs/index.php/online-publication/article/view/435

Reading on a smartphone affects sigh generation, brain activity, and comprehension

https://www.nature.com/articles/s41598-022-05605-0

The Effects of Electronic Books on Pre-Kindergarten-to-Grade 5 Students' Literacy and Language

Outcomes: A Research Synthesis

https://journals.sagepub.com/doi/abs/10.2190/EC.40.1.

Interactive features of E-texts' effects on learning: a systematic review and meta-analysis

https://www.tandfonline.com/doi/full/10.1080/10494820.2021.1943453?scroll=top&needAccess=true

What is the effect of touchscreen technology on young children's learning?: A systematic review

https://link.springer.com/article/10.1007/s10639-021-10816-5

Impact of tablet computers and eBooks on learning practices of law students

https://scholar.google.com/scholar?start=10&q=positive+effects+of+ebooks+and+learning+&hl=en&as_

sdt=0,44#d=gs_qabs&u=%23p%3Dy_WqX52FHQQJ

Nwadialor, Kanayo Louis, and Chukwuemeka Charles Nweke. "ETHIOPIANISM AND SOCIAL ECUMENISM: CHRISTIAN IDEOLOGIES FOR INDEPENDENT MOVEMENTS AND SUSTAINABLE NATIONAL INTEGRATION IN NIGERIA." ETHIOPIANISM AND SOCIAL ECUMENISM: CHRISTIAN IDEOLOGIES FOR INDEPENDENT MOVEMENTS AND

SUSTAINABLE NATIONAL INTEGRATION IN NIGERIA, 3 July 2014, p. 17.

"Overview about Ethiopia - Embassy of Ethiopia." Embassy of Ethiopia -, 7 Dec. 2020, https://ethiopianembassy.org/overview-about-ethiopia/#:~:text=Location%3A%20Ethiopia%20is%20strategically%20located,10th%20largest%20country%20in%20Africa.

"Ethiopian: Search Online Etymology Dictionary." Etymology, https://www.etymonline.com/search?q=ethiopian&ref=searchbar_searchhint.

Bekerie, Ayele. "Ethiopica: Some Historical Reflections on the Origin of the Word Ethiopia." Journal of Ethiopian studies 1 (2016): n. page.

Hatke, George. Aksum and Nubia: Warfare, Commerce, and Political Fictions in Ancient Northeast Africa. NYU Press, 2013.

Khoapa, Bennie A. "Mokone, Mangena Maake (b)." Dictionary of African Christian Biography, https://dacb.org/stories/southafrica/mokone-mangena2/.

Ncurrie. "Richard Allen and the Origins of the AME Church." National Archives and Records

Administration, National Archives and Records Administration, https://rediscovering-black-history.blogs.archives.gov/2021/06/09/richard-allen/.

Adejumobi, Saheed. "Ethiopianism •." •, 15 July 2020, https://www.blackpast.org/global-african-history/ethiopianism/.

Violence in Films and Television

Joanne Cantor, in Encyclopedia of International Media and Communications, 2003

https://openai.com/blog/chatgpt#OpenAI

Chatsonic

https://writesonic.com/chat

www.ingramcontent.com/pod-product-compliance
Lightning Source LLC
Chambersburg PA
CBHW070937180426
43192CB00039B/2313